ALL SMOKE AND MIRRORS

21st CENTURY ILLUSIONS, DELUSIONS, DECEPTIONS, INCOMPETENCE, WILFULNESS, SCAMS, DENIALS AND DOWNRIGHT LIES

BY

D J Haskell

This book is dedicated to my wife Patricia for without her patience and understanding this book would never have been written.

'When men choose not to believe in God, they do not thereafter believe in nothing; they then become capable of believing in anything'

G.K. Chesterton
(1874-1936)

FOREWORD

By

Mark Duchamp

President, Save the Eagles International
Chairman, World Council for Nature
March 2022

Future generations will be wondering why we sacrificed our countryside, our winged fauna and our economy with wind generators that produce an energy that is both expensive and intermittent. May this book, written by David Haskell, help them understand what our fight has been, that of a small critically-minded minority against the unstoppable alliance of ideologues of the global warming theory, unscrupulous businessmen, greedy politicians and scientists looking for funding.

PREFACE

Fear, conspiracy theories, elusions, delusions, illusions, confidence tricksters, fake news, shams, frauds, misrepresentation, rip-offs, scams, swindle, racket, hustle, shakedown, lies, diddles, fabrications or Trojan horse - whatever term or words you may choose, then the meaning would be more than apt for many in offices of responsibility and power in the 21st Century. Indeed, it is a very sad indictment of British governance that many politicians, cabinet ministers, and indeed Prime Ministers do not embrace integrity, honesty and trustworthiness - but as with all things in life there are exceptions that prove the rule. It is the intention of this book to expose and highlight the delusions, ineptness and dishonesty that spin their immoral web through modern society - not wishing to be alarmist, but unless Government changes direction in its many policies, the Nation will be sorely tested.

One of the biggest and disgraceful miscarriages in British history must be the wrongful and dreadful prosecution of many innocent sub-postmasters, when Sir Ed Davey was Post Office minister. It should also be noted that the former Liberal Democratic leader Sir Vince Cable, now retired, was Business Secretary with responsibility for the Post Office at the time, when many prosecutions were taking place. With so many sub-postmasters being accused of theft after the installation of a new Post Office computer system, surely the most cerebrally challenged politician should have smelled a rat, calling for an immediate investigation of the integrity and worthiness of the newly installed computer system. It seems that Post Office bosses were either too dim to do so, or immorally attempted to hide serious errors. There is no acceptable excuse for either the politicians or Post Office management, as at an early stage, the new system appeared to have significant bugs, which could cause the system to misreport; sometimes involving substantial sums of money. It was difficult for sub- postmasters to challenge errors because they were unable to access information about the software to do so. It is unbelievable that so many sub-postmasters were convicted of stealing money, after the Post Office installed its new computer system. Many sub-postmasters pleaded guilty, having felt unable to defend themselves effectively against the resources and intransigence of the Post Office. Many sub-postmasters went to prison. Many became bankrupt, two have since died, one committed suicide. Officials have revealed that more than 400 former postmasters may have been wrongfully convicted over the IT system shambles, and that the number of postmasters wrongfully convicted of theft could be as high as 1,000 the Post Office has admitted. Thankfully, during May 2021, the conviction of theft of 39 former sub-postmasters has been overturned at the Court of Appeal with dozens of former sub-postmasters and postmistresses now being able to follow a clear path in quashing their convictions for fraud, theft and false accounting. This whole affair truly paints the picture of certain politicians being not only incompetent, but deficient in basic morality and honesty as understood by the good citizens of this country.

Margaret Thatcher and Privatisation - now there is a conundrum - did Mrs Thatcher actually have the interests of the general public at heart in her pursuance of privatisation, hoping such a venture would improve living standards, service and drive down prices? This is a question that can be addressed by simply assessing current living standards - have services improved, are prices being driven down, and is the future looking bright as a result of the privatisation of the Utilities? The ever increasing lighting, heating and water bills, the price of a railway ticket, motoring expenses, and the price of general goods should tell their own story. Not only are the public being ripped off with eye watering increases across the board, but savers and their loyalty are also being punished with pathetic returns from saving with banks, building societies, National Savings and yes, less prizes from Premium Bonds! Not to overlook the *sale* of ports, utilities, telecoms, airports, airlines to mainly foreign-owned firms which took away forever annual Government revenues, that kept taxes down. Now all profits are transferred out of the country, and as a result individual taxes have risen.

No doubt a lot of readers have come to their own assessment and contempt regarding Government and its machinations, sleight of hand and downright lies that have been '*served up*' to the public! As an example, it is duplicitous to claim that having the option of changing an electricity company (as a result of privatisation) will drive down costs and afford a better service. Well dear reader, has the cost of your electricity bill decreased, coupled with an improvement in service – exactly! The claim is laughable and disingenuous as moving between electricity companies is simply nothing more than a paper exercise, resulting in changing the name of the company with a new agreement for the Standing Charge and Cost per Unit as seen appropriate by any individual electricity company. It is very noticeable and misleading that the Standing Charge and Cost per Unit are both variable between companies, thus making comparison cumbersome and diminishing clarity. We all depend on the same Power Network (Grid), so why should the Standing Charge differ? It should be combined with the unit cost and simply called a Standard Charge Unit (SCU) resulting in the Same Charge for everyone. No doubt the electricity companies will claim (tongue in cheek) that 'Local Networks' vary and as such justify different charges.

In September 2020, Prime Minister Boris Johnson declared his commitment to make Britain the 'Saudi Arabia' of wind power, as part of the plan to attain 'net-zero carbon' by 2050. Talk about the *Emperor's New Coat*, it is reprehensible that a British Prime Minister is not aware of the simple fact, that if the whole of the UK sunk beneath the waves, then 99 per cent of global emissions will still have to be addressed, such is the total madness of pursuing 'net-zero carbon', whatever that is supposed to mean. Sir Winston Churchill (1874-1965) once said, "Fear is a Reaction. Courage is a Decision." So it is not surprising the 2020/21 coronavirus (Covid-19) pandemic triggered a whole range of conspiracy theories. The slow response of Government, and the unsurprising fatalities of numerous old people being moved from hospital into care homes - it is not difficult to see how conspiracy theories obtain their oxygen. The ruling bodies will surely hope you will NOT read this book as the facts will clearly demonstrate how we are all being manipulated and exploited. It would *appear* that evil *forces* are at work, aided and abetted by immoral *magicians* weaving a sinister web of misinformation. The illusions and delusions are simply that of the *smoke and mirrors* of the typical fairground. The current stupidity that underpins Government policies needs to be halted by wiser heads, leading to more informed, moral and capable representatives being in charge – it is almost as though there is a deliberate and wilful attempt to push Britain back into the distant past - one has to wonder who gains from all these retrograde measures.

It is wise to recollect and ponder on the words by American President Thomas Jefferson (1748-1826)

'The government you elect is the government you deserve'

'I cannot teach anybody anything – I can only make them think'

Socrates, Greek philosopher,
(470 BC – 399 BC)

Contents

'Those who are too smart to engage in politics are punished by being governed by those who are dumber'

Plato
Greek philosopher
(427 BC – 347 BC)

INTRODUCTION

Purposefully the book predominantly deals with the UK electricity generation and supply, for without this vital energy our technological society will collapse. Make no mistake, if the National Grid fails, for whatever reason, it will be truly apocalyptic, thrusting the country back into a new Iron Age and a monumental struggle for those who survive such a catastrophe. If you think I am exaggerating dear reader, then consider everything that you are dependent on for electricity! If you deem civilisation can survive without electricity, then this book will lift the mist and awaken you to the irrefutable truth! Please read on as you could prove instrumental in preventing or minimising a 'Dooms Day' scenario – do not underestimate *People Power* and remember that we all have a political representative to make our views known, and for them to take the necessary action.

It would appear that everything from mild to cold winters, summer heat waves, tsunamis, hurricanes, tornados, earthquakes, flooding, blizzards, melting ice-sheets, glaciers, landslides, to so-called polar bear decline, are all to be blamed on global warming as a consequence of human endeavour, with the industrial revolution as the primary *villain* and cause of increased global temperatures. This global heating is known as Anthropogenic Global Warming (AGW), although nowadays it has *morphed* into what is generally described as Climate Change. This change of wording is seen by the author as duplicitous and manipulative by attempting to *imply* that climate change is something new and *purely* as a result of human activities. Indeed people who challenge the *implication* are seen as climate change deniers and heretics to be metaphorically burnt at the stake, such as Giordano Bruno, the philosopher and scientist, who was burned to death by the Roman Inquisition for his heretical ideas, which he refused to recant in 1600.

However, the planet has always been subject to natural changes in climate as the history of past major and minor ice ages testify. But whether the planet is heading for *catastrophic* global heating has yet to be fully proven. Over the medium to long term geological, chemical and paleontological evidence would suggest planetary cooling and another ice age is the most likely scenario. Nevertheless, please note that any increase in *natural* global average temperatures since pre-industrial times is not challenged by the author, nor is sea level rise since the last ice age. Although it is fully recognised the influence towns and cities have on local temperatures - known as Urban Heat Islands - where the cities and towns are hotter than the surrounding countryside.

It can be argued that the greatest threat to our continued existence on the planet is not directly due to climate change per se, but OVER POPULATION. The Earth has a finite atmosphere, oceanic water, lake, river water and limited land mass - and their bounties are not infinite. Overcrowding, apart from having an influence on the atmosphere, will lead to worldwide pandemics, famine, water shortages, and wars fighting over ever reducing resources – will we be insane enough to wipe ourselves out as a result of a global nuclear conflict?

In an effort to offer a wider perspective to various chapters in this book, the reader should not be surprised to discover that the first chapter deals with climate - it was also felt necessary and helpful for the next chapter to offer a brief history of the power industry – an industry that technological nations now depend upon for their very existence.

It was thus deemed appropriate that the following chapter should start in the home exposing the truth of the so-called Smart Meter. Electrical experience and common sense has been employed to expose Government and the various power company shenanigans and falsehoods. When the electricity companies implore their customers to have a Smart Meter installed, and claim that it will save power consumption, the word *hoodwink* quickly comes to mind. In a sane world can you imagine a butcher, when asked for some lamb chops, advising the customer to visit the grocers and buy vegetables instead; the customer would rightly think the butcher was away with the fairies. So why would the electricity companies wish to install a Smart Meter to cut electricity usage – why would they wish to shoot themselves in the foot - it is illogical and insulting to the intelligence - turkeys do not vote for Christmas! Privatised electricity companies exist to make a profit. If customers use *less* of their product, then without reducing staff numbers, network maintenance, or finding a cheaper source of generation, then surely the only way to make a profit is to push up prices - heads they win and tails you lose. When, dear reader, did you last experience a price reduction in your electricity bill - exactly! Prices will continue to rise. The Government and the power companies are being dishonest in their promotion and pursuance of the so- called Smart meter. It would seem that any system or device defined as *smart* by Government then the opposite is true – are Smart Motorways really smart? Smart meters offer imaginative means of billing, resulting in a potential tariff nightmare that will be dire for all consumers, but lucrative for the power companies.

In the next chapters the book challenges a number of *pseudo green enterprises* such as large scale wind and solar electricity generation, and clearly demonstrates how the public are having the wool pulled over their eyes by an ill-informed and myopic Government. How idiotic that during a video link to a roundtable discussion at the UN in New York, during September 2020, that a British Prime Minister, namely, Boris Johnson said the UK held 'extraordinary potential' for wind energy, and he wants to make a 'big bet' on renewables, turning the UK into the "Saudi Arabia" of wind power. To any informed engineer this is a load of old rubbish - the chapter on wind energy clearly exposes the nonsense and stupidity of the 'big bet' which are truly the words of a charlatan.

Another muddled and inane Government strategy is attempting to generate reliable LARGE amounts of electricity from solar radiation in the UK. Do Ministers not appreciate the latitude and weather in the UK? Solar arrays are not very effective on cloudy days or during the winter months – my four kilowatt roof-mounted solar array will struggle to produce hardly any power on a cloudy day in winter - and of course, the Sun does not shine at night. This nonsense clearly demonstrates that Cabinet Ministers technological and engineering knowledge is truly abysmal, resulting in an appalling strategy to provide an *adequate* and *secure* power supply for the UK. Regarding home-fitted solar arrays I doubt if any Energy Minister, past and present, has a clue that roof-fitted solar panels are in fact *totally depen*dent on a mains supply to enable solar generated electricity to be fed to a house. If the mains power supply fails, so will roof-fitted solar generation as explained in the book. In spite of this however, the reader should be aware there is a good case for home-fitted solar panels, and the chapter on Solar Energy justifies their employment in greater detail; indeed the chapter explains and provides proof of the savings that can be achieved in household energy.

There appears to be no limit to Government shenanigans in its mindless target setting for Electric Vehicles (EVs) with the ban on sales of new petrol and diesel models in 2030, although there are currently no plans for a complete ban on all diesel and petrol cars; the government had originally intended to bring in the ban on new models during 2040. The insights offered in this book will no doubt come as an unpleasant surprise to a number of people who are contemplating the purchase of an electric vehicle. It is not unreasonable to enquire, where all the power will be coming from to supply EVs and additionally account for the intended replacement of conventional boilers with heat pumps. Unlike EVs where it is difficult to estimate how many vehicles will be simultaneously charging, and therefore how much electricity will be needed at any one moment, it is much easier to ascertain for heat pumps. The calculation is made easier simply because during a cold, dark and windless winter's night every household will have a heat pump running to keep warm, that is, apart from empty dwellings - and possibly the odd igloo. During 2017 there were

27,227,700 households in the UK, and the chapter on Alternative Energy offers a meaningful estimation of the additional electricity required, which was found to be of the order of 20 GW for a *minimal* requirement. But heat pumps are certainly not a viable and practical solution to the replacement of conventional boilers as the reader will discover.

A whole chapter is devoted to the folly of generating electricity by Nuclear Fission. It is wise to recall that when the world's first commercial nuclear power station, Calder Hall, was opened on 17 October, 1956 by Queen Elizabeth II, and it was claimed by the media that the electricity generated would be too cheap to measure - but we all know how that turned out. Indeed, at the time of writing the cost for the new Hinkley Point Nuclear Power Station is estimated at £23 billion with a committed cost per unit for 35 years at £92.50 per megawatt hour. The generation of electricity by nuclear is actually the most expensive by far, especially when the full cost of decommissioning is taken into account, which can take up to 100 years for full decommissioning. It should be noted that Germany has set aside £33 billion to decommission 17 nuclear reactors, and the UK Nuclear Decommissioning Authority estimates that the clean-up of UK's 17 nuclear sites will cost between £95 billion to £218 billion over the next 120 years. But who truly knows the time, and total cost to the complete environmental cleaning up of nuclear fission plants? Nuclear fission power stations should not be contemplated by any sane Government as they are horrendously costly and hazardous with the insurmountable problems of decommissioning and disposal of nuclear waste, and of course there is always the possibility/probability of a catastrophic meltdown - we should learn the lesson from the Russians, Japanese and others as highlighted in this chapter.

Later chapters look at the alternative sources of power generation, such as natural gas, tidal and the beaming of electricity (solar energy) from SPACE in the same sense as TV and communication electromagnetic transmissions from satellites. Regarding the use of Interconnectors (cables on the sea-bed) unless Government and the power companies plan to sell electricity to foreign countries (at a profit) there will be no need to build costly interconnectors as the Government should aim to ensure that the UK is *self-sufficient* in power needs. Sensible and controlled fracking would allow this. It should be recognised that fracking has turned the U.S. economy around, and sensible and safe fracking in the UK is explored in the book. To reiterate and stress yet again, Government commitments for UK energy security and dependability are certainly not based on sound engineering, technological, scientific, commercial sense, or indeed that of common sense.

It cannot be *stressed* enough that our 21st Century society very existence depends on a secure and reliable power supply. The Government's insane plans, at the time of writing, could easily lead to the collapse of our technological, computerised, electricity dependent society. It is sad and regrettable the COP26 Conference in Glasgow in November, 2021 turned out to be another flop, when possibly so much could have been achieved. If only countries such as India and China could have agreed to convert their dirty coal-fired power stations to natural gas, there would be a substantial saving in Global CO_2 emissions. Modern gas-fired power stations of the CCGT variety are relatively cheap, quick to build, and efficient offering a 60 per cent saving in CO_2 emissions compared to their coal cousins. Until feasible environmentally friendly means for the generation of large amounts of electricity are developed, then controlled and safe Fracking should be implemented. This will not only make the UK independent on energy, but greatly enhance the economy such that money can be forthcoming for the development of Nuclear Fusion - the Holy Grail of power generation. We are where we are, and Government strategy in the meanwhile should be based predominantly on CCGT power stations, ably supported by tidal and hydro, instead of wasting billions on large scale wind and solar, which by their very nature, are unpredictable and limited sources of power.

It is not just in energy that Government has shown *ineptitude* and *sleight of hand* as there is the *privatisation* of the water companies. Thus in a book of this nature it would be amiss not to examine the Water Company privatisation and who exactly benefits. Under privatisation household water still comes from the same various sources and through the same pipes as before the water companies were privatised – there is not a selection of different company taps in the kitchen to choose from, so where is the competition apart from that of bottled water? It is also very interesting and revealing that Dŵr Cymru (Welsh Water) differs from other water and energy companies by not having any public shareholders - the possible reason for the lack of shareholders is explored in this chapter.

No doubt many readers will be astounded by the range of revelations resulting from Ministerial ineptitude and the hypocrisy - a truly *'Alice through the looking glass'* mad state of affairs. This book has been written for the general public and as such it is recognised that a number of people will not be familiar with some of the electrical units, especially those employed for large quantities. Therefore readers not familiar with such units may benefit from initially referring to Appendix One.

Please have an interesting and enlightening read.

'HE KNOWS nothing; and he thinks he knows everything. That points clearly to a political career'

George Bernard Shaw,
Irish playwright (1856-1950)

CHAPTER ONE

**'It ain't what you don't know that gets you into trouble. It's
what you know for sure that just ain't so'**

Samuel Langhorne Clemens (1835-1910)
(Mark Twain)

CLIMATE

This book challenges a number of so-called *green enterprises* such as large
scale wind and solar electricity generation, thus in doing so, the author
would wish to make it very clear that he is definitely not a climate change
denier - recognising that the Earth's climate has always changed. This
change can occur for many different reasons such as orbital changes, plate
tectonics, volcanism, change in solar radiation and composition of the
atmosphere. We live on a dynamic planet that is subject to many hundreds
of forces and changes, which are instrumental in affecting, directly and
indirectly, not only the weather and climate, but the oceans and land masses.
Therefore it is deemed wise and sensible, before proceeding further, to be
reacquainted with the fundamental geology of the Earth and that of climate.

The Earth has four main layers - the inner core, the outer core, the mantle
and the lithosphere which incorporates the crust. At the centre of the Earth
is the solid inner core, which is mainly composed of iron, and has a radius
of about 1,220 kilometres (760 miles) with a temperature of $6,000^0$ Celsius,
which is similar to the surface of the Sun. The heat of the Earth's solid inner
core heats a liquid outer core composed of iron and nickel, which is about
2,200 kilometres (1,367 miles) thick, and has a temperature ranging from
$4,500^0$ Celsius to $5,500^0$ Celsius. The general consensus of scientists is that
the Earth's magnetic field is generated deep inside our planet. At the centre,
the heat of the Earth's solid inner core churns the liquid outer core. The
churning acts like convection, which generates electric currents and, as a
result, a magnetic field - a natural process

known as a geodynamo - however, it should be noted that there is no way to observe what is happening at the Earth's core.

The magnetic field is like an enormous bar magnet that is strongest near the poles and weakest near the equator. This magnetic field shields most parts of the planet from charged particles that emanate from space, mainly from the Sun. The field deflects the speeding particles toward Earth's Poles. The Sun does not emit the same amount of energy all the time, as there is a constant streaming solar wind and there are also solar storms. The Sun is very stormy, constantly sending out solar flares and high- energy charged particles that travel at speeds of up to 1,609,344 kilometres an hour (1 million miles an hour). The charged particles *excite* gases in the atmosphere, making them glow - just like gas in a fluorescent tube. When these charged particles interact with the Earth's magnetic field and atmosphere ribbons of colourful lights can be seen at the planet's poles. This interaction of particles with gases in our atmosphere results in beautiful displays of light in the sky; oxygen gives off green and red light, nitrogen glows blue and purple. The displays are, or course, the aurora borealis (northern lights) and if you are near the south pole, the aurora australis (southern lights). During one kind of solar storm called a Coronal Mass Ejection (CME), the Sun ejects a huge amount of electrified gas that travels through space at high speeds, resulting in brighter-than-normal northern lights. Under such conditions the aurora may be visible in Scotland and indeed further south - such large storms can be deadly as we shall see later.

The magnetic field is in a constant state of change that waxes and wanes, poles drift and, occasionally, flip. Today the field is about 10 per cent weaker than it was when German mathematician Carl Friedrich Gauss first began measuring it in 1845. Some scientists speculate the field is headed for a reversal - the Earth's magnetic field has flipped many times over the last billion years, according to the geologic record. Reversals take a few thousand years to complete, and during that time the magnetic field does not vanish. It simply becomes more complicated, as magnetic lines of force near Earth's surface become twisted and tangled, and magnetic poles pop up in unaccustomed places. A south magnetic pole might emerge over Africa for instance or a north pole over Tahiti. But it remains a planetary magnetic field, protecting life on the planet from space radiation and solar storms.

At the moment the North Pole is located in northern Canada, about 600 km from the nearest town of Resolute Bay. The last time that Earth's poles flipped in a major reversal was about 780,000 years ago, in what scientists call the Brunhes-Matuyama reversal. The relevant information relating to reversals is captured when molten lava erupts onto Earth's crust and hardens. As the lava solidifies, it creates a record of the orientation of past magnetic fields much like a tape recorder records sound; the alignment much in the way that iron filings placed on a piece of cardboard align themselves to the field of a magnet held beneath it. The reader should be aware that as opposite poles of a magnet attract each other, and the fact that the north pole of a magnet or compass points toward the *Earth's North Magnetic Pole*, one of them has been incorrectly named. Indeed, because opposite magnetic poles attract, then by definition the Earth's north magnetic pole is actually a magnetic south pole and the Earth's south magnetic pole is a magnetic north pole. The direction of magnetic field lines is defined that the lines emerge from the north pole of a magnet and enter into the magnet's south pole. The confusion about the names of the Earth's Magnetic Poles and the poles of a compass or bar magnet are historical and to understand how this incorrect naming came about we must realise the compass was invented in China during the Han Dynasty sometime between 2 BC and 1 AD. These compasses were made of loadstone, a form of the mineral magnetite. The Chinese also developed a naval compass in the form of a magnetic needle that floated in a bowl of water which allowed the needle to stay in a horizontal position. It would appear that as trade between Asia and Europe increased, traders and travellers most likely brought the concept back with them as around the late 1200s to early 1300s, sailors started using a dry compass. Today's compasses have the same basic principle as those from 20 centuries ago, although they are obviously made from more modern materials and have incorporated advanced improvements. For a considerable time, it was thought the compass needle pointed toward the North Star, with that end or pole of the needle being called the north pole of the magnet - often the north pole of the magnet was coloured red and marked 'N'. Thus this naming convention for magnets has been in effect for hundreds of years. It was during the 1600s when astronomer William Gilbert discovered that the Earth had a magnetic field coming from an area near the North and South geographical poles. It was assumed that the Earth was like a huge magnet and those areas were named the Earth's North and South Magnetic Poles. These designations have remained since then and it appears the scientific community felt that it was easier to *say* the Earth's internal magnet had its north pole at the geographic North Pole than to try to change the way compass and magnet poles were designated.

The media are obsessed with global warming and the increase of atmospheric carbon dioxide, overlooking the fact that without carbon dioxide, life as we know it would cease on Earth. But what of the Earth's magnetic field, or more frighteningly, the lack of it! If the magnetic field disappeared or indeed diminished significantly it will have a far greater devastating effect for life on the planet – it is one of the main reasons the Earth can support life. For example, unlike the planet Mars whose atmosphere is less than 1 per cent as dense as the Earth's atmosphere, which is protected by its magnetic field, that repels powerful solar radiation which would not only strip away the atmosphere, but undiminished solar radiation would prove catastrophic for life on the planet. How many readers appreciate there is a Worldwide Network of geomagnetic observatories monitoring changes in Earth's magnetic field. Should we be worried that there has been a general decrease in the strength over the past hundred years - today it is about 10 per cent weaker than it was when German mathematician Carl Friedrich Gauss started observations in 1845. NASA claims the magnetic shield is now weakening at 5 per cent per decade. Scientists cannot say with any certainty whether this represents a fluctuation or whether it is a decrease which will eventually lead to a reversal – the uncertainty could be that the magnetic poles are preparing to flip and they just do not know? Although in order for a reversal to take place it is claimed there must be a brief time during which the field is non-existent – during this *brief period* it begs the question of what effects will undiminished solar radiation have on the planet? According to Earth's geologic record, the planet's magnetic field reverses, on average, about once every 200,000 years, and the time between reversals varies widely. However, as mentioned earlier, the last time Earth's magnetic field flipped was about 780,000 years ago, so I guess we are well overdue for another reversal.

A large discharge from the Sun such as a CME, as mentioned earlier, can have a profound effect upon the Earth and would have the potential to be devastating to a modern highly technological society, which is heavily dependent on computer systems and electricity. It is very sobering to realise that it was not that long ago that such an event took place; during September, 1859 a powerful CME hit the planet's magnetosphere and induced the largest geomagnetic storm on record. This storm was known as The Carrington Event which triggered strong aurora displays and caused havoc with telegraph systems with operators receiving electrical shocks. Frighteningly if such a storm occurred today it would have the potential to cause widespread panic. There would be a very significant disruption of the Grid with resulting blackouts, as a consequence of damage to switching gear and transformers, not to mention satellite and computer systems failing, plunging society into chaos. How many readers know a solar storm that occurred on 23rd July, 2012 was in fact a large and strong CME, which luckily missed the Earth with a margin of approximately nine days! Next time we may not be so lucky - although one event that does not bear thinking about, and that is if a large CME hit the Earth at the very same time when the magnetosphere is diminished!

For readers who may be interested the Sun is a 'pussy cat' when compared to the next nearest star to the Earth, that is Proxima Centauri at a distance of 4.3 light-years. Solar flares from this star are 100 times greater than the most powerful of the Sun. A person standing on the surface of one of the two known exoplanets that orbit Proxima Centauri would be exposed to such flares, not once in a century, but possibly several during a day.

The next layer is the mantle and is the Earth's thickest layer. It lies between Earth's dense, super-heated core and its thin outer layer, the crust. The mantle is about 2,900 kilometres (1,802 miles) thick, and makes up a whopping 84 per cent of Earth's total volume. The upper mantle extends from the crust to a depth of about 410 kilometres (255 miles). The upper mantle is mostly solid, but its more malleable regions contribute to tectonic activity.

Finally, the lithosphere is the solid outer section of Earth, which includes Earth's crust, as well as the underlying cool, dense, and rigid upper part of the mantle. The lithosphere extends from the surface of Earth to a depth of about 70-100 kilometres (44-62 miles), whilst the crust is between 5-8 kilometres (3-5 miles) deep under the oceans (oceanic crust) and about 40 kilometres (25 miles) deep under the land masses (continental crust). The crust itself is made up of what is known as Tectonic Plates which are in constant motion with volcanoes and earthquakes found at plate boundaries – such as the *Ring of Fire* which is a region around much of the rim of the Pacific Ocean, where many volcanic eruptions and earthquakes occur. The Ring of Fire can be described as a horseshoe-shaped belt about 40,000 kilometres (24,855 miles) long and up to about 500 kilometres (311 miles) wide. The tectonic plates are divided into 7 major and 8 minor plates. The largest plates are the Antarctic, Eurasian, and North American plates. Plate tectonics is the theory that the Earth's crust and upper mantle are composed of numerous major and minor plates that fit together tightly but are in continuous motion, moving sometimes toward one another and at other times apart. The movement is known as plate motion or tectonic shift, and it has been going on for a long, long time. According a latest study the Earth's plates are moving faster now than at any point in the last 2 billion years.

Fossils found in Antarctica support tectonic shift and that this frozen wilderness has not always been located at the South Pole. It is claimed that in the distant geological past Antarctica was sited across the equator, and approximately 450 million years ago the crust that makes up England was in the Southern Hemisphere. Indeed, about 300 million years ago all the Earth's continents consisted of one large mass called Pangea and centred on the equator. This large land mass was surrounded by a super-ocean called Panthalassa. During the *Triassic* period, about 200 million years ago, this super land mass broke up into two smaller supercontinents, called Laurasia and Gondwanaland, which were separated by the Tethys Sea. Both Laurasia and Gondwanaland then later broke up into the present day continents some 66 to 30 million years ago. As a point of interest in the 1950s Professor Charles Hapgood, a historian, believed that Antarctica was not covered with ice 11,600 years ago, and his *cataclysmic pole shift hypothesis* suggests that there have been geologically rapid shifts in the relative positions of the modern-day geographic locations of the poles and the axis of rotation of the Earth, creating calamities such as floods and tectonic events. Alas these subjects are beyond the scope of this book, but are recommended reading.

Regarding the Earth's climate, it is important to understand that the oceans, atmosphere, land and vegetation are all part of the climate system. This is not to be confused with weather which is the condition of the atmosphere over a short period of time, whereas climate is the atmospheric conditions over relatively long periods of time. Therefore climate is a description of the average weather conditions in a certain place for the past thirty or so years. It should also be recognised that different areas of the planet have different climates. Easy examples are the polar and equatorial regions. Climate is influenced by lots of different things, such as how near or far it is from the Equator, its proximity from the sea, and the elevation of the ground. Changes in various aspects of the climate system, such as the size of ice sheets, the type and distribution of vegetation, the temperature of the atmosphere or oceans will influence the large-scale nature and features of the atmosphere and oceans. Climate has changed in the past, whether there has been humans or not, and will continue to do so into the future. History has shown that in the Northern Hemisphere there was a warm period from approximately 250 BC to 400 AD, then a period of cooling. This was later followed by the Medieval Warming Period occurring from the 10th to the 14th century, a period when the Vikings settled in Greenland and parts of North America. This latter warm periods was followed by what is known as the Little Ice Age, which occurred roughly between 1300 AD and 1850 AD, where canals and rivers in Europe were frequently frozen deep enough to support ice skating and winter festivals. London was no stranger to the many frost fairs on the River Thames. Weather is always changeable in Britain and has, overtime, experienced its fair share of storms.

We are all familiar with the author Daniel Defoe and his book *Robinson Crusoe*, but how many are aware of his book *The Storm*, which described what was called the worst storm in history, which killed thousands of people and resulted in hundreds of thousands of pounds worth of damage.

During a terrible night in November, 1703 Britain experienced an extreme weather event. After weeks of winds and rain, a cyclone blew through the land covering an area from the Welsh coast to the Midlands and the south of England. It is claimed between 8,000 and 15,000 lives were lost and the lead roofing was blown off Westminster Abbey. The high winds caused in the region of 2,000 chimney stacks to collapse in London and 4,000 oaks were lost in the New Forest. It is remembered through history as the 'Great Storm of 1703' and is indeed a contender for the worst storm ever recorded in Britain.

The *Great Storm* of 1953 was also said to be Britain's worst peacetime disaster, and with no severe flood warnings in place and telephone lines down, people were completely unaware of the devastation which was about to hit. In Britain the storm claimed the lives of over 300 people, although in the Netherlands some 400,000 acres flooded, causing at least 1,800 deaths and widespread property damage. The 1953 North Sea flood was a major event caused by a heavy storm at the end of Saturday, 31st January 1953 and the morning of the next day. The storm surge struck the Netherlands, north-west Belgium, England and Scotland. The combination of gale-force winds, high spring tide, and low pressure caused a tidal surge, which caused the North Sea to rise up to five metres above its average level. It was forecast the storm would not happen more than once in 250 years and it was the reason that the Thames Barrier was built.

But then there was another *Great Storm* of 15th October, 1987 making a sudden appearance in the middle of the night. It was one of the UK's most catastrophic cyclones, leaving 18 people dead and a huge repair bill of more than £1 billion; much of southern England was completely battered by the winds, gusting up to 161 kilometres per hour (100 mph). This was the result of a severe depression in the Bay of Biscay that had moved northeast. Among the most damaged areas were Greater London, the East Anglian coast, the Home Counties, Channel Islands and the west of Brittany. Despite warnings on French TV news, the BBC TV weatherman, Michael Fish, infamously failed to warn Britain of the devastating storm, with the dismissive words: "Apparently a lady rang the BBC and said she heard that there was a hurricane on the way. Well, don't worry, if you're watching, there isn't." I can still clearly remember Michael Fish and his TV weather forecast, also the subsequent damage quite vividly when we visited the wife's auntie in Brighton not long after the storm.

Apart from such storms and floods there was the winter of 1962–63, known as the Big Freeze of 1963, which was one of the coldest winters (December, January and February) on record in Britain. Temperatures plummeted and lakes and rivers froze over. The Arctic weather did not relent until March, 1963. Central England experienced its coldest winter since 1740. Rivers, lakes, and even in places the sea, froze over. Snow continued to fall during the month of February, which was stormy with winds reaching gale-force, and a 36-hour blizzard caused heavy drifting snow in most parts of the country. I have first-hand experience of the harsh conditions in 1962/63 as a telephone and overhead line maintenance engineer during that period. Such weather events are not easily forgotten, and I can still recollect driving down snow-covered country lanes to repair overhead *bare* wire (copper cadmium) telephone lines. It was quite a challenge and it was akin to driving through snow tunnels with the top of the tunnel removed, such that it was necessary to 'double-up' in the vans due to the extreme conditions. If the maintenance vehicle became stuck on ice, there was little room to open the van doors to get out and push in high- sided country lanes. As young engineers, we actually enjoyed the challenge coupled with a sense of achievement in repairing emergency lines such as those lines connecting country doctors, and the farms that could be reached. In those days, due to line shortage, there existed telephone connections known as party lines (shared service), where two customers, via a joint connection on the local telephone pole, shared the same line back to the exchange. Of course, only one customer could use the line at any one time. To identify which telephone was calling the exchange the customer had to press a button on the telephone. This button connected an earth (ground) condition to one of the conductors of the telephone line. As the telephone line consisted of two conductors each party line caller had their dedicated conductor for earth connection enabling correct identification. The actual connection to earth was via a rod driven into the ground to establish an electrical path back to the exchange. Unfortunately during the Big Freeze (1963) the ground became so frozen and hard that it shrank away from the earth rod diminishing its effectiveness, resulting in the customer being unable to call the exchange. I can still see the customer's puzzled look when asking them for a kettle of hot water to pour around the earth rod. This procedure enabled effective (electrical) contact with the ground, before a deeper (frost free) and more effective earth rod could be driven into the ground.

Although I can barely remember the event, except for snow almost to the top of the house front door, the winter of 1947 was also very harsh in Britain. Starting on 23rd January, Britain experienced several cold spells that brought large drifts of snow to the country, blocking roads and railways, which caused problems transporting coal to the power stations, resulting in massive disruptions of the electricity supply to homes, offices and factories. People suffered from the persistent cold, and many businesses shut down temporarily; animal herds froze or starved to death.

Towards the end of February, fears of a food shortage began as supplies were cut off, and vegetables were frozen into the ground. All this resulted in severe hardships in economic terms and living conditions as the country was still recovering from the Second World War. Then the middle of March brought warmer air to the country which thawed the snow lying on the ground. This snowmelt rapidly ran off the frozen ground into rivers and caused widespread flooding.

Regarding global climate the planet has experienced major ice ages and there have been at least five major ice ages. Scientists have identified past ice ages by studying ice cores, deep sea sediments, fossils, and landforms. Ice and sediment cores reveal an impressive detailed history of global climate. Nearly 3 billion years ago there was the Huronian ice age (or Makganyene ice age (2.4 billion to 2.1 billion years ago). This was followed by the Cryogenian (720 million to 635 million years ago), then the Andean-Saharan (450 million to 420 million years ago), and the Late Paleozoic (335 million to 260 million years ago). Lastly the Quaternary, which started 2.7 million years ago and ended about 11,500 years ago. During this period, the climate repeatedly changed between very cold periods, during which glaciers covered large parts of the world, and very warm periods during which many of the glaciers melted. The cold periods are known as *glacials* (ice covering) and the warm periods are called *interglacials;* the planet is currently in an interglacial period called the Holocene - during the Earth's history climate repeatedly changed between very cold periods, and very warm periods. There have been at least 18 cycles between glacial and interglacial periods, and it should be noted that glacial periods lasted longer than the interglacial periods. The last glacial period began about 100,000 years ago and lasted until 25,000 years ago - as mentioned above today we are in a warm interglacial period. But how long will this warm period last? No scientist knows for sure, except to offer that the last four interglacials lasted for approximately 20,000 years – thus it would appear we have about another 10,000 years before investing in fur coats - for those readers who require further information on interglacials then researching the paleoclimate record for the Devils Hole, Nevada, U.S. and the Vostok ice core from Antarctica should satisfy this need. The disciples of Anthropogenic Global Warming (AGW) claim that man-made carbon dioxide is responsible for atmospheric warming. I would suggest this addition to the atmosphere might be the cause for some increase in global warming, but it is certainly not proven to be the main driving force regardless to what the media and a number of climatologists may claim. It is interesting that AGW has morphed into the words *climate change*, which suggests obfuscation as climate change can also mean global cooling as we have seen with glacials.

There are many significant forces and mechanisms that are responsible for climate change. After the freezing winter of 1962-3 there was a lot of discussion in the media about another ice age – yet now it appears we are going to cook? Political shenanigans and pseudo-science obfuscation lend themselves to a lot of confusion, The waters are further muddied with numerous scare stories such as the BBC asking 'If snow will become a thing of the past due to climate change' in 2016. Unbelievably during 2007 Welsh scientists were warning that the principality's highest mountain would be snow-free in 13 year's time (2020). Well what a foolish prediction, and it comes as no great surprise that snow was found at 1500 ft on Snowdon on 1st January, 2021. Then on the 5th of January heavier falls were experienced generally above 1000 ft. To be sure, during April, 2021 a light sprinkling of snow was seen at on the Carneddau, Snowdon, and at 1000 ft near Ogwen at 0900 GMT on the 5th of the month. These are facts that can be easily verified. There are claims that the sea ice in the Arctic is decreasing, thus the Northwest Passage will be open all through the year, coupled with stories in the media of polar bears drowning due to lack of ice. The truth is that Polar bears can swim incredible distances such that polar bears regularly swim over 48 kilometres (30 miles). In October, 2019 the Polar Bear Specialist Group of the International Union for the Conservation of Nature (IUCN) released a new assessment of polar bears, and the findings revealed the most up-to- date data for polar bear populations: Although most of the world's 19 populations have returned to healthy numbers, there are differences between them. Some are stable; some seem to be increasing, with 4 populations in decline due to various pressures.

There is only a narrow zone that supports life on Earth and it is called the Biosphere. It is limited to the waters of our planet, a fraction of the crust, and the lower regions of the atmosphere. The Biosphere, by its very nature, is a dynamic system subject to various influences which change atmospheric, sea and land temperatures over time. The climate of the Earth changes over the short, medium and long term. Short term climate changes are obviously the seasons, such as in the northern hemisphere, experiencing spring, summer, autumn and winter; these changes being due to the inclination of the Earth's axis and orbit around the Sun. Indeed, it may be argued, with tongue in cheek, that a very short climate change is the difference in temperature between day and night in, for example, a desert region – extremely hot during the day and very cold at night. Long term examples of climate change, as previously mentioned, are the major ice-ages, with the last major ice age reaching its maximum about 20,000 years ago and ending about 11,500 years ago.

The forces which contribute to weather, the short and long term climate change, are many and extremely complex, and there are a number of known mechanisms causing this. Indeed, a theory by the Serbian geophysicist and astronomer Milutin Milankovitch (1879-1958), which is now known as the Milankovitch Theory; states that as the Earth travels through space around the Sun, cyclical variations in three elements of Earth-Sun geometry combine to produce variations in the amount of solar energy that reaches Earth.

Firstly, there is the Earth's tilt (known to astronomers as the obliquity of the ecliptic), which ranges from 24.6 degrees to 22.1 degrees over a 41,000 year cycle; the current axial tilt is about 23.44 degrees.

Next, there is the shape of the Earth's orbit around the Sun which is determined by its eccentricity. Where the closer the eccentricity is to zero (0) the closer the shape is to a circle: the eccentricity of the Earth is 0.0167 and the periodicity of changing from less elliptical to more elliptical is about 93,000 years.

Thirdly, there is the change in the direction of the Earth's axis of rotation, which is known as Precession. The axis of rotation behaves like a spinning top and as such traces a circle on the celestial sphere over a period of time - a wobble of the Earth's axis taking almost 26,000 years for each cycle. Currently the axis points towards Polaris (the North Star), but in 13,000 years' time it will point towards Vega. This is the brightest star in the constellation Lyra, the fifth brightest star in the night sky and the second brightest star in the northern celestial hemisphere. It is a relatively close star at only 25 light-years from Earth; Vega was the northern pole star around 12,000 BCE.

According to Milankovitch his theory explains why the Earth cycles in and out of glacials and interglacials.

But of course, there are many other forces that affect the weather and climate, which in no particular order of merit are the major ocean currents such as the Gulf Stream, Labrador and North Equatorial Current etc. Air currents that give rise to Global Jet Streams, which are streams of fast flowing, narrow, meandering air currents in the atmosphere. Usually the jet stream above the UK marks the boundary between the cold polar air to the north and warmer air to south. It is called the Polar Jet and plays a significant role in the weather across Britain. Other forces being the El Niño Effect, La Niña Effect, Sunspot Cycle, Cloud Formation and Cosmic Rays, Dansgaard-Oeschger Cycle, Volcanoes, and other long term influences such as the passage of the Solar System (passing through dust clouds) around the Galactic centre, and as mentioned earlier plate tectonics whereby the land masses move over the surface of the planet from warm to cold areas, and vice versa, altering ocean currents etc.

Weather and climate change are truly complex, and it is difficult to understand the forces of nature and the impact that human society has on the planet. To claim that our modern society cannot have any effect upon the large environment we call Earth is being perhaps a little naïve. The use of an analogy is helpful such as imagining a bath-tub full of fresh water into which a grain of salt is introduced. This single grain will have a very minute effect and the bath water will not taste salty. But if additional grains of salt are put into the bath water, at say, the rate of one grain a minute, then at some stage a *detectable* change will take place in the bathwater's salinity. The point to note is that there will indeed come a *time* when a change is significant; this *time* being directly related to our sense of taste, the volume of water, size and salinity of the grain of salt, and the rate of input - thus given time, small changes can have a profound effect. The oceans cover most of the planet, being vast and deep and containing almost innumerable life forms. But who would have imagined that the vast numbers of cod could be devastated by over-fishing in the waters of the Grand banks, off the coasts of Labrador and Newfoundland. The destruction of the Grand Banks cod is one of the biggest fisheries disasters of all time. Surely the Portuguese and Basque fishermen who worked the Grand Banks as early as the 15th century would have deemed it impossible
– the world's richest fishing grounds – tragically owing to human failing (greed, lack of foresight and conservation) it has happened! Closer to home in the North Sea it was thought that cod stocks were limitless, but over-fishing of the North Sea Cod stock resulted in a near collapse in 2001, although recent reports suggest there may be a slight recovery – time will tell! Bluefin tuna, which shoaled the Atlantic in vast numbers, were thought to have declined by 50 per cent during the 1960s, but now the figure has dropped to nearer 80 per cent. World-wide many other fish stocks that have been over-fished - the oceans may be vast but their resources are not in any way infinite.

Similar events have happened on land and the decimation of the North American Bison (buffalo) is a good example. Millions of wild buffalo once roamed the North American Continent from Mexico to Canada, long before people settled there. The coming of the Native Americans who relied on the buffalo for practically everything from food and clothing to shelter did not have any meaningful impact on buffalo numbers. But later when the European settlers arrived in America things began to change as the Native Americans learned to use horses, thus expanding their hunting range and enabling more buffalo to be killed. But it was only when white trappers and traders introduced guns that the killing rate began to spiral resulting in the slaughter of millions of buffalo. Unbelievably even train passengers were shooting buffalo for sport during the 19th Century, and I can still remember reading comics as a small boy and being enthralled by the adventurous tales of Buffalo Bill (William Cody) never realising this man, who was hired to kill buffalo, slaughtered more than 4,000 buffalo in two years – such is the innocence of children. Currently I believe there are between 150,000 and 200,000 bison throughout North America, although the vast majority of them are raised on ranches for commercial purposes (mostly for meat, hides and skulls). It is shameful to recognise that there is not a single area, from pole to pole, whereby human intrusion, exploitation or contamination can be denied. It is truly outrageous that nearly 400 marine species are at risk due to the tons of plastic shopping bags, fishing nets and other waste dumped in the seas every year. Due to apathy, ignorance and selfishness the Loggerhead Turtle, North Atlantic Right Whale, Hawaiian Monk Seal, Shearwater, African Penguin and unbelievably the beautiful Albatross could be wiped out as a direct consequence of *our* pollution. So it should come as no great surprise that we do have the ability to affect the survival of species, not just locally, but even on a global scale.

Most of us are aware that there is a temperature difference between town and country. It is warmer in the cities than the countryside in both summer and winter – the urban heat island effect. All the concrete, roads and buildings of an urban environment absorb solar radiation and give out heat. It is useful to remember that electrical night storage heaters convert electrical energy to heat energy, by taking advantage of cheap rate night-time electricity. This energy is used to warm up bricks inside the storage heater, and then during the day this stored heat is released at a controlled rate to warm rooms. Thus a city can be seen, in this sense, to be a very large storage heater, but of course there is no containment and control knob to regulate the amount of heat radiated. The city absorbs and radiates heat during the day as city dwellers know only too well, but cities and towns also radiate heat at night. The warming of urban environments by solar radiation is further enhanced by the artificial heat being generated in all the buildings from lighting, although much less with Light Emitting Diode (LED) lighting. The warmth is increased during the winter months owing to central heating and other forms of heating. Additionally, population density must also be taken into account as mammals generate heat, and of course there is heating from the internal combustion engine. All these sources of heat contribute to what are known as *Urban Heat Islands*. It can be argued that a small number of these Urban Heat Islands will have little, if no effect on global temperatures. But what has to be considered is how the spread, enlargement and linking up of many urban heat islands will it take before they do indeed begin to have a significant effect, not only on inter-urban island temperatures, but maybe on a global scale. Additional warming will be beneficial up to a point, but just in the same context as light pollution it needs control and containment as any astronomer will tell you. Are you able to see the heavens and the multitude of stars from where you live dear reader, and have you ever seen the beautiful Milky Way from your urban heat/light island? I doubt it very much, and you only have to look at a night-time satellite images of the Earth to see how light pollution has spread dramatically across the face of the planet. It should be recognised though that many historical temperature readings will come from areas before they were urbanised, leading to misleading claims on warming in many areas.

With regard to the reliability of predictions, during April, 2009 the Meteorological Office (MET) stated the UK was odds-on for a barbecue summer. But this turned out to be an inaccurate forecast as July and August turned into washouts. Then during the same year forecasters said we could expect a milder than average winter, with only a one in seven chance of a cold Christmas season – in the event it turned out to be the coldest and snowiest winter for more than three decades. During March, 2012 the Met Office predicted 'drier' than average conditions for April to June, and the quarter turned out to be the wettest since 1910 with widespread flooding. The Met Office has now stopped issuing long-term forecasts to the public and instead it continues to give 'probability' guidance for coming months to Government departments such as the department for Environment, Food and Rural Affairs (Defra) which needs to plan ahead. It is not good science to claim global warming is *definitely* caused by man-made emissions and that there is a massive body of evidence supporting this claim. There may be a substantial consensus for this assertion in parts of the media, and political arena, but there is certainly not a massive body of scientific evidence to support the 'anthropogenic' warming hypothesis as yet - there are many other more significant forces at work that can explain climate change. It is interesting that after 94 years of weather forecasting for the BBC, the state-owned Met Office has been replaced (August, 2015) by the private weather company MeteoGroup; I will offer no comment and leave the reader to make of this what they will.

Historically an increase in atmospheric carbon dioxide has FOLLOWED warming, and not the other way around. It is a fact that as the oceans heat up they release carbon dioxide. Rising temperatures make carbon dioxide leak from the oceans for two main reasons. First, melting sea ice increases the rate that the ocean mixes, which dredges up carbon dioxide rich deep ocean waters, and secondly, when the oceans are *warmed* up, it is similar, in a way, to *warming* a bottle of Coke or Soda - it drives the gas out! But, of course, if man-made carbon dioxide is in fact driving up *natural* global atmospheric warming, then we are indeed in trouble, as this will only exacerbate the situation. Natural increases in carbon dioxide concentrations have periodically warmed Earth's temperature during ice age cycles over the past million years. The warm episodes (interglacials) began with a small increase in amount of solar energy our planet receives due to small changes in Earth's axis of rotation or in the path of its orbit around the Sun. During the last 650,000 years there have been seven cycles of glacial advance and retreat, with the end of the last ice age about 11,700 years ago marking the beginning of known human civilization. Nevertheless it is recognised that air bubbles trapped in mile-thick ice cores (and other paleoclimate evidence), it is has been found that during the ice age cycles of the past million years or so, carbon dioxide never exceeded 300 parts per million (pm) - before the Industrial Revolution started in the mid-1700s, the global average amount of carbon dioxide was about 280 ppm. According to the National Aeronautics and Space Administration (NASA), the amount of carbon dioxide (CO_2) in Earth's atmosphere was about 416 parts per million (ppm) in April 2021, and CO_2 has been rising for 200 years.

The disciples of man-made global warming should fully recognise and admit that the natural forces and mechanisms that drive the climate have been in play long before the Industrial Revolution, or indeed the appearance of Homo sapien; they need to also be aware of the relatively new field of science called Cosmosclimatology. Sensible climate scientists admit that the Earth's climate is determined by hugely complex systems, and reliable prediction is extremely difficult, if not impossible. I am not claiming Humans do not have the capacity to have an impact on global temperature – I am simply saying that scientists just do not know *for sure* if global temperatures, *in the long term, are on the increase and are directly the result of human activities*. During 2000 Dr David Viner, head of the climate unit at the University of East Anglia, said that in future: "Children just aren't going to know what snow is." Remember that none other than Al Gore stated that the Arctic would likely be ice-free in summer by the year 2014. I wonder what their opinions are now in 2022 – are they scraping egg from their faces as a result of such deluded prophecy?

It was enlightening to read in the press during November, 2021 that the Government Met Office predicted a mild winter, whereas the BBC MeteoGroup Weather Service predicted the weather to be cold and harsh. Well they both cannot be correct, and it is instructive, given their multi-million pound computers, satellites, weather balloons and ships, etc, weather forecasters still struggle to accurately predict the weather in a couple of months time. Thus how much faith can we put in such organisations when they attempt to predict how climate will change? The reader should be aware that whilst the Met Office is publicly funded, MeteoGroup is a private weather forecasting organisation. During 2018 the BBC Weather Forecast service changed its data supplier from the publicly funded (Government) Met Office, to the privately owned MeteoGroup. Then during 2019 MeteoGroup was integrated with the U.S. weather services company Data Transmission Network and Dataline (DTN).This is a private company that is based in Burnsville, Minnesota, USA. The joint companies will be known as DTN and have a European headquarters in Utrecht, Netherlands. The integration of MeteoGroup with DTN will result in 200 meteorologists working in weather rooms across the planet, coupled with the addition of more than 10,000 weather observation stations, resulting in the largest network of private weather observations worldwide, which should result in more accurate weather forecasting – time will tell.

Regarding global warming, as in all things, it is very important to recognise that consensus is not a scientific principle. The word consensus belongs to the world of politics and not science. A consensus is general agreement among a group of people. In a democracy, if the general consensus is that a certain party is the favourite and as a consequence has the most votes, then that party should be elected. But if the consensus is that the Sun orbits the Earth (the geocentric model) as in the days of Aristotle, then that is not true science, as it is simply a case of what many people consider the truth. The Greek and Roman civilisations believed in the geometric model that the Sun, Moon, planets and stars all orbited the Earth. But as history has demonstrated the believers in the geometric model were completely wrong as science and the truth have prevailed. It should also be recognised that Isaac Newton's classical theory of gravitational force held sway from his Principia, published in 1687, until Einstein's General Theory of Relativity in 1915 - but Newton's theory is sufficient even today for all but the most precise applications – after all it enabled man to land on the Moon.

In September/October 2021, during her foolish diatribe to a seemingly gullible audience at the Youth Climate Summit in Milan, Italy, the young Greta Thunberg enigmatically branded the British Government as Historical Climate Villains. But such is the *naivety* of youth this misguided young woman appeared not to be aware that the UK, nowadays, is responsible for **less than one per cent of global emissions!** If, by her definition, the UK is a Climate Villain, then what words does she have for far more polluting countries such as China, India, USA, and Brazil? Such is the worldly wisdom of one so young, she opined that science does not lie, but fails to understand that consensus, in ignorance, can! Indeed, science can be guilty of untruths due to *ignorance* of all the facts. A very simple example was the commonly accepted consensus that the Milky Way Galaxy was the Universe – that is, before the American astronomer Edwin Hubble came along. Prior to Hubble, the Milky Way Galaxy was believed to be the entire Universe. Which, of course, to modern day science is not true, and was hence an unwitting lie.

In concluding this chapter on climate and a *'wetting of the appetite'* to the following chapters the following is a brief insight to the delusions and how politicians, the energy and water companies, are *'pulling the wool over the public's eyes'*.

They say a good salesperson can sell snow to an Eskimo and sand to an Arab, hence it follows the growth of the unbelievable market for bottled water in the UK, especially when it can be argued that Britain should have the best and healthiest tap water in the world, owing to the annual rain fall and safe drinking regulations. So what motivates people to drink so-called 'healthy' water from a plastic bottle – do they not trust the water companies? It is disgraceful that discarded plastic bottles are not only threatening land based wildlife, but are breaking down in the oceans and being ingested by all manner of marine life. This pollution and poisoning of marine life, will in turn, affect many people who eat the bounty of the seas. Thus it is totally irrational to see so many folk on the street clutching their bottles of water – a first-class lesson in the power of persuasion and clever marketing, highlighting the naivety and gullibility of many people. But what of the power companies with their ugly and ineffective wind farms and limited solar parks, which are industrialising and despoiling our irreplaceable and glorious countryside. The English landscape artist, John Constable (1776-1837) would have been horrified at such desecration to the beauty of the British landscape. Indeed, wind farms and solar parks in such places as the English countryside, across the wild beauty of Wales and Scotland surely must be classified as an act of mindless vandalism.

In Wales there is the despoliation of areas such as the enchanting Brechfa Forest (which was the ancient Glyn Cothi Forest) with two wind farms namely Brechfa Forest East comprising 12 wind generators with a capacity of 24 MW, and Brechfa Forest West with 28 wind generators and a capacity of 57.4 MW, giving a total of just 81.4 MW. This output is derisible when compared to the Pembrokeshire 2000 MW CCGT Power Station which can, if necessary, power the whole of Wales 24 hours a day, recognising that when there is little or too high a wind then Brechfa Wind Farms will produce zero electricity. Such is the utter madness in the provision of these wind farms that thousands of trees were cut down for the sites. The forest used to cover some 6,500 hectares and is looked after by National Resources Wales (NRW) for the benefit of people, wildlife and timber production. Please note the forest is supposedly for the benefit of people and wildlife. Many people objected to these wind generator schemes and it is well known that wind farms kill wild birds and bats. Thus how could Carmarthenshire County Council Planning Committee justify their approved planning application for Brechfa Forest East in their meeting on 17th December, 2013? It was surely a betrayal of local people, wildlife, forestry and common sense; goodness knows what tourists and visitors make of it all.

During 2019 an independent review into how NRW handled timber contracts has found 'serious failings'. Auditors Grant Thornton found problems were so bad they 'heightened exposure to the risk of fraud'. The agency which is responsible for publicly-owned woodland across Wales, repeatedly sold its timber without going to the open market. NRW said it was taking the findings 'very seriously' and had an action plan to ensure improvements were made. The Welsh Government said they were satisfied NRW was 'taking a strong lead on putting things right in their organisation'. The way timber was sold had already been heavily criticised by auditors and the assembly's public accounts committee. NRW called in experts, Grant Thornton, to review its processes after the Wales Audit Office put a black mark against its 2017/2018 accounts for the third year running. Surely readers will not be too impressed with how NRW has managed its timber responsibilities, not to mention the building of wind farms which are not exactly compatible with many people's wishes, wildlife and forestry.

Many good people who have little understanding of how the power industry functions, have been hoodwinked through clever marketing to the false notion of how wonderfully effective, and environmentally-friendly wind farms are, whilst not realising that it takes thousands of wind generators to replicate the same power output of a single, large sized, conventional power station. How many readers realise that 4,000 large 2 MW, 123 metre (400 feet) high wind generators will be needed to replace a 2000 MW power station such as the Pembrokeshire CCGT gas-fired power station. Of course, these numerous wind generators would not be adding a single kilowatt of power to the Grid when there is little or too much wind – such is the deviousness of developers and the madness of politicians. Indeed, how many people actually appreciate that unbelievably, Scottish wind farms are paid millions when they are not producing any power for the Grid when being shut down to avoid damage from high winds - these wind farms would not be out of place in Alice's Wonderland, and as a consequence of an energy policy recommended at the Mad Hatter's Tea Party. Attempting to generate large amounts of power from pseudo-renewable sources such as wind farms will result in not only tears at bedtime, but cold hands and feet as well.

Another ludicrous means to generate large amounts of electricity is the construction of *industrial size* solar parks in the UK, especially at our latitude and with our variable and unpredictable weather. People coming home on a cold, cloudy and dark winter's evening will be really annoyed to find that there is no power available for heating or lighting, if they are fully reliant on a large industrialised solar park. Being connected to both a large wind farm and solar park will still not guarantee useful power, as during high pressure in winter there can be little useful wind, thus people will then be both disappointed and annoyed. You do not have to be the sharpest pencil in the box to realise that Government policy for UK Energy in the 21st Century requires a total reassessment, or the lights will indeed be going out. The billions being wasted on Wind Farms and Solar Parks should have been channelled to tidal and ocean current electrical generation, especially considering the UK is an island, with the second highest tidal range on the planet in the Bristol Channel. At least 5 per cent of total UK electricity could be generated from a barrage in the Bristol Channel, in addition there is potential to further generation from tidal lagoons, such as the Swansea Bay Tidal Lagoon. Elsewhere around the British coast there are other suitable locations for tidal energy. Currently and in the foreseeable future the dominant generation should come from natural gas, the balance from other sources such as tidal and hydro as explained later. Talk about not being able to see the wood for the trees - it would seem that none of our leaders have any knowledge or understanding relating to science, technology or engineering!

CHAPTER TWO

'When the Paris Exhibition (1878) closed, electric light will close with it, and no more will be heard of it'

Oxford Professor Erasmus Wilson (1809-1884)

BRIEF HISTORY OF ELECTRICITY

Having discussed climate, it will be useful to briefly to look at the history of the Alternating Current (AC) power producing industry we are familiar with in the UK today. I wonder how many people know that the world's first central electrical power station produced Direct Current (DC) and was built in the United States, by Charles Francis Brush (1849-1929). Charles Brush designed and developed an electric arc lighting system that was adopted throughout the United States and abroad during the 1880's. It should be noted the arc light preceded Edison's and Swan's incandescent light bulb in commercial use and was very suited to places where a bright light was required, such as street lighting and illuminating public and commercial buildings. A fundamental requirement in Brush's arc lighting system was the dynamo (electric generator). This generator was the workhorse of the centralised power station - a concept developed independently by Brush and Edison and which eventually evolved into the electrical power generating industry we know today. Prior to Edison, Charles Brush had a centralised power station operating in New York City during 1881. Thomas Edison's centralised power station did not materialise until 1882 and was situated in Pearl Street, lower Manhattan, New York City. It started producing electricity during September of that year illuminating Wall Street offices including those of the New York Times. All the early power stations generated DC, and they were small and unable to distribute electrical energy over any appreciable distance due to the loss of energy in the line conductors in the form of heat, known as power loss. When electricity passes through a metallic line conductor it creates heat due to the resistance of the conductors and thus energy is lost. This power loss is mathematically proportional to the square of the current, times the resistance of the conductor. Therefore to transmit a useful current over a large distance would require metallic line conductors of large diameters which are very expensive, thus making direct current transmission over large distances uneconomical. Due to his marketing techniques, Edison soon became a dominant presence in the early central station electrical industry. Edison realised that doubling the output of a steam-powered generator did not double its capital cost, and a larger steam engine was more fuel-efficient, generating more electricity from a given amount of fuel. Although Edison could obtain a large steam engine, he had to design a generator that was compatible to it! He also had to construct cables sufficiently durable and effectively insulated to carry the currents over long distances - currents much greater than those used in telegraph or telephone cables. Edison also realised that the many thousands of lamps operating off his network of wires could not be connected in series, for the very obvious reason that if one lamp was switched off or failed, then current would cease to flow in the network rendering all the other lamps useless. He realised they all had to be connected in parallel, which in itself presented a problem - with a low resistance element lamp, as more and more lights were switched on, then more and more current would flow in the network - this would demand very heavy (low resistance) and expensive cables, if the current were not to melt them. The way out of this dilemma was resolved when Edison and his staff, at their Menlo Park laboratory, during 1879, produced a carbonised bamboo fibre in an air-free globe - the incandescent bulb - this had a high resistance which presented Edison the breakthrough to push ahead with his large-scale central station electricity system.

It should be noted the invention of the incandescent bulb by Edison in America was challenged by Joseph Swan of Newcastle-upon-Tyne, England, who demonstrated a very similar design. Nevertheless the invention of the incandescent bulb at more or less at the same time, greatly hastened the demise of the electric arc light. The early electric light bulbs used a filament of carbon drawn out to a very fine wire and enclosed in a bulb which had all the air extracted. The decision to use carbon was simply that it could be heated to a higher temperature, than any of the metals known at the time, and exclusion of the air prevented the filament burning away. Unfortunately, the temperature was limited and the carbon lamp gave off a very reddish light; during the early 1900's certain rare metals, having a high melting point, were used in an evacuated bulb. Then in 1913 the Tungsten filament bulb was developed. This used a gas-filled bulb filled with an inert gas such as argon, into which was placed a Tungsten filament which could run at a temperature of about $3,500^0$ K. This bulb emitted a light nearer to daylight than any of the previous lamps, although still containing too many red waves and not enough of the green and blue; additionally, the proportion of light waves to heat waves is greater in this type of lamp, so that more light is obtained for a given amount of power than in the earlier types. The amount of light emitted by a lamp is measured in 'lumens' and a small gas-filled lamp gives approximately 16 lumens of light for every watt of electrical power which it consumes. Therefore an ordinary 60-watt light bulb used for domestic lighting will give out about, 960 lumens of light.

The early central generating stations were in effect, analogous to a gas-making plant. The generating station fed electrical current through a 'loop' of wire extending for miles to customers premises, to which electrical lamps would be connected. These wires, as mentioned above, would be similar to telegraph cables, although being much more robust, because of the heavy currents they had to carry and maintain. Thus this 'electrical mains' and electrical lighting was analogous to the 'gas mains' and gas lighting at the time. A major disadvantage with the early central generating stations is that they all produced direct current, that is, current that flowed in one direction only. Although the direction of current flow was not that critical, the problem was due to power loss in the cables - as the system expanded and the wires extended further and further, the losses from the wires became more and more expensive! The use of heavier and heavier wire was not the answer as they also became more and more expensive. So the direct current system had its limitations - but an answer was found in the use of alternating current electricity. With the advent of the transformer (1831) it was found this new device could raise or lower the voltage of an alternating current as required. An electric wire carrying a given amount of power, the higher the voltage, the lower the current, and therefore the lower the losses in the wire. This simply means that if an electrical generator fed the primary winding of a suitable transformer with 7,500 volts at 32 amps, then the secondary winding could be designed to offer 132,000 volts at 1.8 amps. If the secondary winding of this transformer is then connected to the transmission network the line losses are reduced. Therefore an AC generator at a power station has its voltage output stepped up by a transformer - this stepped up voltage is then transmitted over the network wires with minimal power losses - at the far end the voltage is then stepped down, for customer usage. Thus an AC system, due to its minimal power losses, will cover a substantially larger area than a DC system. It also meant that electricity could be generated at remote sites where water was freely available and could be employed to produce what is known as HYDROELECTRICITY and indeed, in the early days most electricity was produced this way and in many parts of the United States electricity and was commonly known as 'hydro'. The provision of high-voltage electricity became known as TRANSMISSION so as to distinguish from the relatively low-voltage electricity that was fed to customers, this low-voltage being called DISTRIBUTION. George

Westinghouse of the United States was a leading pioneer in the use of AC and was one of Edison's most challenging rivals. During the 1880s there was fierce competition between the DC and AC systems, each having its various merits. The DC system was relatively simple and not so costly for customers close together in city centres. Whereas AC was more economical for customers spread over a large area in the suburbs and rural areas. Additionally, in those early days DC could run motors whereas AC could not. It was during the 1880's when 'three-phase' was introduced - which reduced the number and size of wires - that the AC system had another cost cutting boost. Thus with the advent of the transformer offering the ability to transmit electrical energy over large distances, *resulted in large central alternating current power stations becoming economically viable.* The growing success of the AC system meant that hydroelectricity became very popular, especially in the United States, due to the fact that there were plenty of water resources. One of the most famous engineering projects in the United States was the construction of a large hydroelectric dam across the Colorado River, known as the Hoover (Boulder) Dam.

It is interesting that the world's first DC hydroelectric power scheme was British and developed by Sir William George Armstrong, an engineer and industrialist who founded the Armstrong Whitworth manufacturing company on Tyneside. The company built ships, locomotives, automobiles, armaments and aircraft, and it was a major British manufacturing company of the early years of the 20th century. Together with Richard Norman Shaw, the renowned Victorian architect, Armstrong built the country house Cragside, Northumberland. During 1870, a Siemens dynamo was installed, creating the world's first domestic hydroelectric plant, which was used to power Cragside and the estate farm buildings. During a 2006 regeneration project extensive rewiring was carried out and a later date (2014) a new screw turbine with a 17-metre (56 ft) Archimedes' screw was installed. It has a capacity of 12 kW which can supply 10 per cent of the properties electricity consumption. It should be noted though that the world's first AC commercial hydroelectric power scheme was American. In 1895, using the work of Nikola Tesla, the industrialist George Westinghouse built the first hydroelectric power plant at Niagara Falls, New York. The plant generated power for nearby Buffalo, only 35 km (22 miles) distant from the power plant. This achievement beat Tesla's rival Thomas Edison, and was the final victory for Tesla's *alternating current* over Edison's DC hydroelectric power station, which was built on the Fox River in Appleton, Wisconsin, USA. The station, later named the Appleton Edison Light Company, was initiated by Appleton paper manufacturer H.J. Rogers.

Nikola Tesla (1856-1943)

It would be appropriate in concluding this chapter to address the misunderstanding and lack of recognition regarding the development of the electricity industry, and how history has been unkind to the electrical genius that was Nikola Tesla. Praise has been heaped upon Thomas Edison, George Westinghouse and indeed Marconi, when much wider acknowledgement should be given to Nikola Tesla. He was a man way ahead of his time and who brought about numerous technological breakthroughs by inventing, amongst other things, alternating current, neon light, the induction motor, radio, shadowgraph (x-rays), and indeed a radio-controlled boat. Unfortunately, the full range of Tesla's genius and achievements are beyond the scope of this book, although I would recommend a book titled, 'THE MAN WHO INVENTED THE TWENTIETH CENTURY' by Robert Lomas. This is a book that will surely help in putting Tesla's extraordinary achievements back on the pages of electrical history, a man who deserves full recognition and not to be so widely forgotten.

Make a REALLY DUMB decision...

...Have a SMART METER Fitted!

CHAPTER THREE

Non semper ea suntquae videntur

(Things are not always what they appear to be)

SMART METERS

A word of caution! If you have already *elected* and have had a Smart Meter fitted then perhaps you should skip this chapter, as it will only result in a feeling of foolishness for being taken in by all the *hype*. There is also a certain amount of **reiteration** in this chapter, the reason for which is to fully emphasise the misleading and fraudulent claims perpetrated by Government and the power companies.

It can be argued that the *promotion* of the Smart Meter is a perfect example in deceiving and confusing people; the campaign by 'Smart Energy GB' can be compared to the art of fairground tricks that employ smoke and mirrors to fool people. During March, 2019 the Advertising Standards Authority (ASA) ruled that a radio commercial for Smart Energy GB was misleading, because the meters themselves do not deliver any reduction in electricity use. The commercial was banned and Smart Energy GB was told to make clear that any claimed energy savings are dependent on households monitoring their energy use and cutting back.

Smart Energy GB is a not-for-profit, government-backed campaign which claims to help everyone in Britain to understand the importance of Smart Meters and their benefits to people and the environment - the national campaign is aimed at homes and small businesses across England, Scotland and Wales. They say that, if you have not got a Smart Meter yet, you are missing out – Really! From the consumers point of view, the heading for this chapter is misleading and it should more accurately read, *The Not-so-Smart Meter*. Although, it could be defined as a S*mart Meter* for the power companies as it affords the *potential* control over supply and 'imaginative' billing of customer usage. Indeed, if for whatever reason, the power company feels the need to isolate a customer, the meter offers the *potential* to remotely switch off the power at customer's premises without a power company employee such as an engineer calling to the property. However, a Department for Business, Energy and Industrial Strategy (BEIS) spokesperson has said: "Network companies cannot remotely 'turn off' smart meters, nor could they control the amount of energy supplied to homes without the express consent of consumers. Any proposals from network companies to do this would be rigorously challenged by Ofgem which serves to protect consumers." It should be noted that it is misleading to claim the customer needs a Smart Meter to *monitor* power usage and cost. The claim is inaccurate as a separate MONITOR is needed for this requirement, as we shall see later. It would appear that any system or device defined as S*mart* by Government then the opposite is true. Therefore, before discussing the misleading and disingenuous concept of a Smart Meter, please forgive a slight digression, as I feel it will be instructive in recognising how Government wilfully misuses words and their true meaning, with the following observations pertaining to the so- called Smart Motorway.

Certain parts of the UK motorways are becoming prime examples to the misuse of the word smart. Ministers, at the time of writing, effectively have their heads in the sand and will not listen to intelligent dialogue, facts, or take any sensible action in their blind support of the so-called *Smart Motorway*. It is misleading and ingenuous to claim that Smart Motorways are designed to ease UK motorway congestion, by permitting cars to be driven on the hard shoulder at least some of the time, with traffic being monitored via cameras and 'active' speed signs which can vary the limit. But official figures reveal the fatality rate on so-called Smart Motorways is up to a third higher than that of conventional highways with hard shoulders. It has been reported that deaths on such roads have surged as more have been rolled out across the country, while on normal motorways rates have fallen - despite Smart Motorways being dubbed the 'safest roads in the country'. Speaking to a BBC Panorama investigation during 2020, The Rt Hon Grant Shapps MP said: *"We absolutely have to have these as safe or safer than regular motorways, or we shouldn't have them at all."* (The Rt Hon Grant Shapps was appointed Secretary of State for Transport on 24 July, 2019). But this is in sharp contrast to the views of Automobile Association (AA) and the Police Federation. On a BBC Panorama programme during January, 2020 the AA president Edmund King was very blunt and said they were dangerous and not fit for purpose. This is something he has been saying for years, and what makes the matter worse is it seems that the technology on which they are based, namely radars and detectors to identify vehicles in distress are not all they are cracked up to be. It takes an average of 34 minutes for emergency or rescue services to reach a stranded vehicle, and as a consequence, the unfortunate driver is a sitting duck for more than half an hour. John Apter of the Police Federation told the BBC Panorama programme that they were informed the technology behind Smart Motorways would be so advanced, it would detect obstructions, it would detect problems on the motorway instantaneously, but the technology is not there. A Freedom of Information request submitted about one stretch of the M25 found that near-misses had increased 20-fold since the motorway was made 'Smart' unbelievably from 72 in the five years before the hard shoulders were removed to 1,485 in the subsequent five years as a Smart Motorway. On 28th January, 2020 a former transport minister said that he was effectively misled by the Highways Agency when he gave approval for the smart motorway network to be rolled out. Hopefully, by the time this book appears in print, sanity will have prevailed and the concept of so-called Smart Motorways will be abandoned.

Returning to the concept of the so-called *Smart Meter* it would be foolish to ignore the delusion that they are for the customer's benefit - Smart Meters are being introduced for the medium and long-term benefits of the power companies. The installation of the Smart Meter has, so far, cost an eye-watering £11 billion, and guess who pays? Yes, it is you and I dear reader in our power bills. It is not surprising energy bills keep on rising, and a considerable number of homes have yet to be fitted, so there is still a couple of billion to be spent on installations – happy days! With a conventional electricity meter a representative (engineer) from the electricity company has to visit the household, if it is necessary, for whatever reason, to cut off the electricity supply. But the Smart Meters can offer a means to isolate any customer or customers if the Grid, for example, falls below standard owing to lack of generation. Smart Meters also offer imaginative means of billing, resulting in a potential tariff nightmare that will be dire for all consumers – a consequence of muddled and mule-headed policy makers. Please note dear reader that all observations in this chapter relate only to the electricity Smart Meter and readers are left to their own research, assessment and conclusions relating to the Gas Smart Meter.

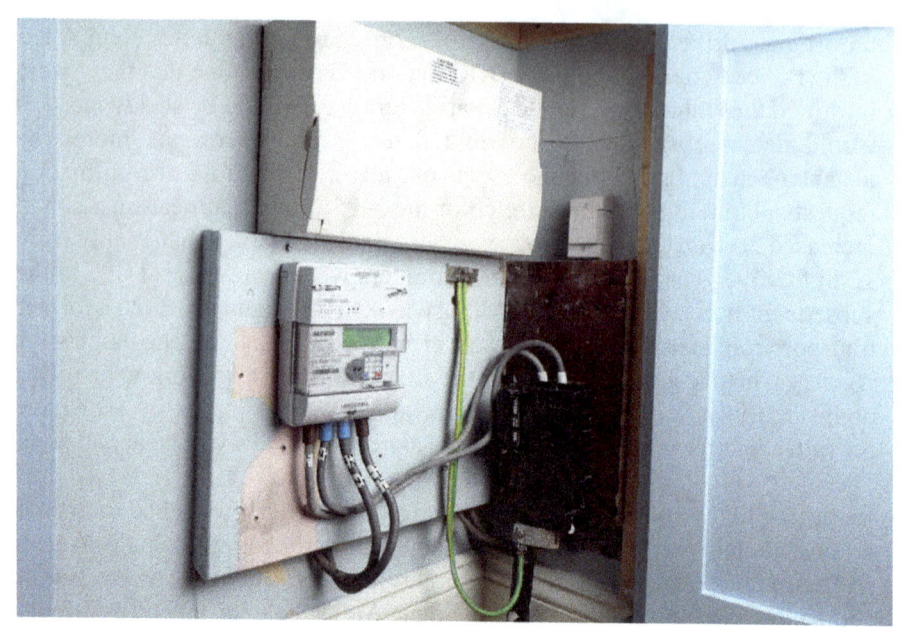

MAINS TERMINATION,
SMART METER AND CONSUMER UNIT

The electricity Smart Meter is an electronic device that records customer electrical consumption and communicates the information to the electricity supplier for monitoring and eventual billing. The Smart Meter has the ability to record energy hourly or more frequently, and will facilitate two- way communications between the meter and the central system. The communications from the meter to the network can be wireless, or via fixed wired connections such as the power line. In promoting the Smart Meter, Robert Cheesewright, Director of Policy and Communications at Smart Energy GB, patronisingly informs the public that switching off the light, will save energy and that a Smart Meter will help in this endeavour. This is insulting as we all have access to a far superior means in making such decisions – it is called a brain. Robert Cheesewright can also be accused of, *sleight of hand*, as a Smart Meter alone, as mentioned earlier, will NOT show how much energy is being used at any one instance in terms of money; a separate MONITOR is required. According to the Government Department for Business, Energy and Industrial Strategy (BEI), it was claimed in 2018 that 1.5 million Smart Meters installed in homes were not working properly. Indeed, many Smart Meters stopped working when customers switched supplier and customers had to go back to submitting their own readings. Over 900,000 Smart Meters were installed in homes but were, in effect, 'dumb' and doing nothing more than a traditional meter. Additionally over 600,000 were not working because they had been put in new-build properties that had not yet been occupied, or the customer had switched to a small supplier and the Government did not have up-to-date data – thus a total of 1.5 million meters were not operating in smart mode. During 2018 households reported more than 3,000 Smart Meter-related problems to the Citizens Advice during 2017. This resulted in the consumer group calling on the Government to delay the roll-out by three years. One of the most common complaints was that the Smart Meter stopped working when having switched supplier. Other complaints encompassed were problems when having a Smart Meter installed, with engineers failing to turn up or being unable to fit the device into the space available. Some were also frustrated that despite having a Smart Meter they still had to submit readings manually. During July, 2018 MPs called on the Government to urgently review the Smart Meter roll-out which they said was "Over time, over budget and mismanaged." The British Infrastructure Group of MPs and Lords warned that half of Smart Meters stop working when customers switch supplier, while a tenth were not functioning due to poor mobile phone signal.

The Director of Policy and Communications at Smart Energy GB, Robert Cheesewright, must have had his tongue in his cheek when he said: "By getting a Smart Meter, we can all save money on our gas and electricity bills and stop manual reading and inaccurate estimated bills. The Smart Meter system is so much better than our outdated analogue system, which is why Government wants suppliers to ensure every household is offered a Smart Meter by 2020." But as already mentioned it is disingenuous to claim a Smart Meter alone can save money on bills as the saving is completely dependent on the customer reading a separate monitor and taking the appropriate action. There appears to be no end to the insulting smoke screens as a Smart Meter advert in a national newspaper on Monday 19th October, 2020 had a person stating, "The display tells me exactly how much energy I'm using pretty much as I use it. I can see what I'm spending as I boil the kettle or turn off lights in an empty room." This is extremely misleading, and again to reiterate and stress, that a Smart Meter alone cannot do this as a separate monitor (DISPLAY) is needed. As mentioned earlier, it is extraordinary that every household, that is, you and I dear reader, whether we want a Smart Meter or not, will be forced to fork out £420 to help fund the £11 billion Smart Meter project. The cash being taken from our energy bills - just imagine what £11 billion would do for the NHS or repairing roads. It beggars belief that Government advises folk to change energy companies to save on bills when Smart Meters are not even compatible between the various energy companies. No doubt Oliver Hardy would have said, "That's another fine mess you've gotten me into Stanley!" The previous Secretaries of State for Energy and Climate Change such as Chris Huhne, Edward Davey and Amber Rudd have a lot to answer for. Amber Rudd, for example, read history at Edinburgh University and with the greatest respect, one has to wonder as to what engineering or technological expertise did she bring to the table - could this be one of the reasons why the Government is in such an unbelievable mess regarding UK energy strategy?

When it comes to understanding power consumption in the home most people will recognise that it is far cheaper to power a five watt LED lamp bulb than say, a three thousand watt electric fire. To be sure, for the same period of use as the electric fire, six hundred LED lamp bulbs could be powered for the same cost - it is not rocket science. Do we need or have the inclination to continually stare at a device to see how much power is being consumed hour by hour. If this rocks your boat then there are a number of monitoring devices available on the market, and as such there is absolutely no need for a Smart Meter! The Government backed organisation, Smart Energy GB, (www.smartenergygb.org/en) as this book reveals, has numerous *questionable* adverts appearing in the press. The following is a quote from one of their many adverts, '*Millions of people are also looking to Smart Meters, which provide near real-time energy readings, helping consumers to identify the areas in their homes where they might be able to turn off unnecessary gadgets and appliances, or find other ways to cut back on consumption, it's a daily means of working towards a greener environment*'. End of quote.

TYPICAL SMART METER DISPLAY UNIT (MONITOR)

To reiterate, Smart Energy GB state on their website that, 'Smart Meters are the new generation of gas and electricity meters being rolled out across Great Britain. They show you how much energy you are using in pounds and pence'. But Smart Energy GB adverts are not fully informative and in fact disingenuous since they do not make it clear that when having a Smart Meter installed it is usual to be provided with TWO electrical devices. There is the SMART METER itself connected at the interface of the electricity network and the house wiring. Then there is the in-house display unit (IHD) for use by the customer. The in-house display unit is usually known as a MONITOR. It is this device that will show the amount and cost of energy being used. Thus the Smart Meter by itself will not offer any more information than a conventional analogue or digital electricity meter. Thus it will be necessary to have an additional monitor in the house to show how much energy is being used at any one instance in both kilowatt hours and money. It is vitally important to realise a separate monitor should be part of the Smart Meter package.

Electricity display monitor

As has been explained a Smart Meter alone will not display the combination of how much energy is being used, and what it is costing at any one instant. For those folk who have a conventional meter fitted and wish to see their usage in kilowatt-hours and cost, then a MONITOR can be obtained separately without having to agree to have a Smart Meter fitted. These devices have been on the market for many years. During January, 2014 I purchased an Efergy Elite wireless portable electricity monitor that displayed the amount of electrical energy a household consumes at the time the display is read, also giving a reading showing usage in financial terms. I should point out that a search on Internet sites such as Amazon will show the many different makes of electricity monitors that are available over a varying price range. Initially, I was full of enthusiasm in observing and confirming the various responses between lighting and high consumption appliances such as an electric fire, or an electric oven. None of the readings though came as a surprise as my wife and I fully appreciate the difference in consumption between low energy demanding devices such as lighting and high energy consuming heating appliances. Unsurprisingly it was not long before the enthusiasm wore off, and looking at the display quickly became a boring and time wasting exercise. I had far better things to do than continuously consult a monitoring device. The display certainly did not encourage the switching off any electrical appliances as my wife and I were already very diligent in our energy consumption, especially with the high cost of electricity these days. Without question Smart Energy GB are misleading the public by failing to offer the full picture and explain that you do NOT need a Smart Meter if you wish to monitor your electricity usage and see how much it is costing.

Therefore if the reader wishes to monitor their electricity usage the answer is simply purchase an electricity monitor. The Efergy Elite wireless electricity monitor cost £39.99 back in January, 2014 and I would recommend the device, even though I seldom consult it nowadays. I do recognise though, that a lot of folk will find a monitor useful, especially those who do not have an electrical background or who are unfamiliar with electrical units. The Elite Classic Electricity Monitor is a low cost, quick and simple to install monitor. It is an easy to use device that will help the customer to understand domestic electricity consumption. The monitor indicates the energy usage and cost in real-time. As the device is wireless and thus portable, the user can walk around the house and observe the impact of various electrical devices by switching various appliances on and off. The Elite Monitor affords, portability and compact design, historical data (day, week and month), tracks carbon emissions, shows room temperature and humidity, can view average energy consumption, audio alert indication of power excess, multi-tariff ready, holds up to 24- month data, and multi-currency (£, €, $, cP, RKr). The Efergy website states that the 'Which Magazine' has consistently ranked the Elite Classic Electricity Monitor as the best energy monitor in the UK having compared it with other 15 different energy monitors in the market.

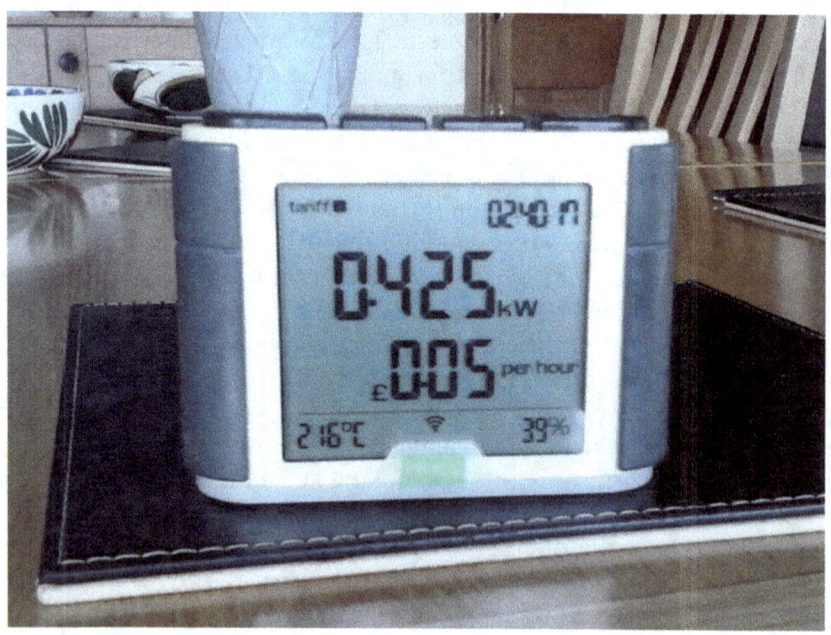

EFERGY ELITE WIRELESS ELECTRICITY MONITOR

Regarding the claim of the power companies (and others) to the saving of energy and money by having a Smart Meter fitted, the truth is easily unravelled by considering a simple analogy of traders in the high street, such as the butcher, the baker and the candlestick maker. All these traders obviously wish to sell their products and make a profit – that is what they do! They certainly would not encourage customers to purchase less of their offerings, resulting in minimum, or no sales. Rightly or wrongly, the world of business motivates itself on profit. So we have a conundrum with the power companies as it can be argued that this is exactly what they are attempting by persuading their customers to have a Smart Meter fitted - assisting and encouraging the consumer to use less electricity. But what if everyone, as a result of having a Smart Meter fitted, used less electricity, then the income of the power companies would obviously drop and thus their profits, something their shareholders might not be too happy with. Remember power companies are no longer state controlled having been privatised. Surely the mission of all private companies is to show a profit and thus provide a dividend for their shareholders. So who is kidding who?

In one of their advertising features, Smart Energy GB stated that according to Dr Stephen Hall, researcher, University of Leeds, Smart Meters can enable Electric Vehicle (EV) owners to be even more environmentally friendly, by matching charging with the greenest electricity on the system. Putting EV and Smart Meters together offers an incredible prize, and sustainable driving. Well, dear reader, what exactly does *matching charging with the greenest electricity on the system* actually mean? A Smart Meter is simply a device for measuring electrical power wherever the incoming generated electricity originates. There is absolutely no method to differentiate between different electrical energy sources at the Smart Meter or indeed at a conventional electricity meter. This would necessitate a direct (dedicated) power line to either a wind farm or solar park for example, to ensure a green electricity supply. If the reader wishes to acquaint themselves with the various types of generation that contributes to the National Grid at any one moment, a visit to the G.B. National Grid Status website, (www.gridwatch.co.uk) will show the various sources and contribution in *real* time. No doubt this access to power generation knowledge is something that Smart Energy GB, Government and the Power Companies would prefer you not to know – a website where you can actually engage with the truth. As an example, on Friday, 11th January, 2019 at 13:15 GMT UK total demand was 44.03 GW of which coal generation provided 3.44 GW (7.8%), Nuclear 6.46 GW (14.67%), CCGT 24.22 (55.01%), Wind 3.27 GW (7.43%) the rest being made up from Biomass, Hydro, OCGT, Pumped Storage, *Estimated Solar*, and French, Dutch, and Irish Interconnectors. For people not familiar with the term Interconnector, they are basically sea-bed cables through which the UK imports or exports electricity, see later in this book. For comparison at 10.30 GMT on 27th October, 2020, a wet and very windy day, UK total demand was 36.9 GW with gas contributing 13.39 GW, wind 8.29GW and solar a mere 0.81 GW, the balance coming from other sources.

Billing

The electricity companies claim that a Smart Meter will give the customer accurate bills, but I would challenge this claim, as past experience has indicated otherwise. I have a habit of sending a meter reading to my power company on the same day every month and they still cannot *get the bill right* with three consecutive months being in credit. For the three summer months of 2018 my bill was in credit by £23.72 for June, £10.05 for July and £21.51 for August, 2018. Interestingly my Direct Debit to this power company was £41 for June, £45 for both July and August, which begs the question of why did they deduct £45 for two months when it should have obviously been reduced. No doubt a power company will claim that since a Smart Meter offers *continuous* data being sent, rather than monthly they will be on top of things and therefore be able to issue bills more accurately. As mentioned earlier the Smart Meter offers the power companies the ability to be very imaginative with their billing such as an hourly and variable unit charge - you can bet your bottom dollar it is the electricity companies that will come out on top, and not the consumer – ask yourself when you last had a reduction in the cost of electricity. Remember shareholders desire greater and greater profits - if not, there is little motivation in purchasing electricity shares.

Remote disconnect

As mentioned above Smart Meters not only lend themselves to imaginative and exploitive billing, but will have the potential to isolate a customer or group of customers from a remote computer. We should all be very concerned at this scenario, as at the moment mains electricity disconnection requires a visit to the home by a representative of the electricity company. Traditionally, utilities send a metering service person to isolate the home by disconnecting the main fuse. Failing this, a power engineer has the option of disconnecting the power line drop-off if supplied by overhead cables, or at the nearest joint if fed by underground feed, although this would be a very time consuming exercise, and in a number of cases quite difficult.

As a comparison, the reader should be aware that to disconnect a landline telephone, for whatever reason, will not require a visit by a telephone engineer. All landline telephones have a dedicated pair of wires back to the telephone exchange and can be isolated at the local exchange, without a visit to the home by an engineer. Although older readers will remember shared service lines where two properties were connected together on the local drop-off pole and shared a single pair of wires back to the exchange. In this case disconnection of a particular subscriber would be carried out on the telephone pole. With regard to mains supplied electricity, homes are directly connected to the local power network and do not have a dedicated pair of wires back to the power station or sub-station. This means, as mentioned above, that to isolate a single home would require the visit of a metering service person or engineer, which is both costly and time consuming.

The advent of the Smart Meter offers electricity companies the potential to remotely disconnect, so isolation of one or more customers will not become a problem and indeed whole areas could potentially be cut off if the power companies deemed it necessary. I cannot speak for everyone, but I can survive without a landline telephone, although life would become very difficult without mains supplied electricity. The Citizens Advice will inform you that if you have a Smart Meter, your supplier could potentially disconnect your supply remotely without needing access to your meter. But, before they do this, they must contact the customer to discuss options for repaying debt, for example, through a repayment plan, or visit the home to assess the personal situation and whether this would affect you being disconnected, for instance if you are disabled or elderly. If the supplier does not do this and they try and disconnect you, then a complaint to the supplier should be made. Citizens Advice will also inform you that a Smart Meter will not automatically save you money – you will have to be proactive to reduce your energy costs. The best way to do this is to use the digital 'in-home' display that is offered with a Smart Meter to keep track of how much energy is being used, and then trying to reduce it - but after reading this chapter dear reader, you will understand that it is not necessary to have a *Smart Meter* fitted. However, a Department for Business, Energy and Industrial Strategy (BEIS) spokesperson has said: "Network companies cannot remotely "turn off" smart meters, nor could they control the amount of energy supplied to homes without the express consent of consumers."Any proposals from network companies to do this would be rigorously challenged by Ofgem which serves to protect consumers."

I wonder how many readers are aware of the term Advanced Metering Infrastructure (AMI) which is the collective term to describe the whole infrastructure from Smart Meter to two-way communication network to control centre equipment and all the applications that enable the gathering and transfer of energy usage information in near real-time. Such an advanced metering infrastructure differs from automatic meter reading in that it offers two-way communication between the meter and the supplier. The aim of an AMI is to provide electricity companies with real-time data about power consumption, allowing customers to make informed choices about energy usage based on the price at the time of use. I would suggest that under this system the customer can look forward to some very imaginative and perhaps confusing billing. Customer systems include in-home displays, home area networks, energy management systems, and other customer-side-of-the-meter equipment that enable smart grid functions in residential, commercial, and industrial facilities. The reader should also be aware that a broadband connection is not required for Smart Meters as they transmit information on a separate wireless network via Global System for Mobile Communications (GSM) mobile networks. The meters use low power radio frequency signals to collect and transmit information relating to services such as electricity, water and gas. The meter reading data is collected and collated at access points and forwarded to the power company over the existing mobile networks in the same way as a call or text is sent – Smart Meters contain a GSM SIM card. For the technically minded the Smart Meter RF transmitter is typically in the 902 MHz and 2.4 GHz bands. Power output is typically of the order of 1 watt in the 902 MHz band and much less in the 2.4 GHz band. Currently there are two main types of smart meters, with the older model known as the SMETS 1 (Smart Meter Equipment Technical Specifications). Unfortunately this version of the meter has an Achilles Heel, in that if the user changed their electricity supplier then it was likely the meter would not work to the new supplier, and became a 'dumb' meter for operational purposes. However most of these are currently being upgraded, which will allow smooth switching between suppliers. The latest type of Smart Meter is known as SMETS 2, which became available in 2018, and has a more advanced specification using a more modern communications network. All electricity suppliers will now use the new meter enabling smooth transfer between electricity companies.

This sounds all hunky-dory, but the mobile network operators such as EE, 02, Vodaphone and Three have committed, with Government backing, to the phasing out 2G and 3G technologies by 2033. During December, 2021 the Rt Hon Nadine Dorries MP, Secretary of State for Digital, Culture, Media and Sport, announced that the government has agreed with the UK Mobile Network Operators that 2033 will be the date by which all public second generation (2G) and third generation (3G) networks in the UK will be switched off, promising a further £50 million to boost the performance and security of UK 5G networks. The 'upgrade' for 2G and 3G General

Packet Radio Service (GPRS) networks to 4G and 5G, will enable faster data transfer speed. Thus it should come as *no surprise* that the current Smart Meters will become *obsolete* when the 'mobile' signals change, as the meters will not be compatible with the new 4G and 5G technology. Yes, you could not make it up! A technician will be needed to visit households to either change the 'communications hub' in the meter, or completely change the meter. It can be argued that fundamentally, as a cost to the country and customer, any savings in fitting a Smart Meter, is completely and utterly swallowed up in the £11 billion already spent!

Summary

Changing from one electricity company to another, an engineer will definitely not turn up at the customer's premises to connect the house to a *different supplier*, or change the *meter*. The house or building will still be connected to the same property feed, whether it is overhead or underground, and to the same local network, switching and generation plant, with absolutely no guarantee of being fed by so-called environmentally-friendly electricity, In fact, it is dishonest to promise total 'Green' energy, unless the consumer is connected *directly* to a Green Energy Generation source, which, of course, the consumer is not! It is outrageous for an electricity company to advertise 'Total Green Energy' when the National Grid cannot differentiate between sources of generation for feeding to customer's premises. The electricity companies should be more open and concede that Grid energy is a mix from more than one source of generation. How many people fully appreciate that changing electricity companies is simply nothing less than a paper exercise – the whole concept is designed to manipulate and exploit the customer. To reiterate, the electricity companies are there to sell their product – it is what they do. Even the most cerebral challenged in the village will realise that any company not selling sufficient electricity will soon be in trouble, unless subsidised by another body such as Government, or unwittingly by the customer. Shareholders will not be happy bunnies if customers, as a consequence of having a Smart Meter fitted use less and less electricity thus driving down profits - unless, of course, shareholders are saved by the electricity company hiking its prices to the customer - how is your power bill faring lately dear reader?

The various carrots being trotted out for having a Smart Meter installed do not stand up to scrutiny, especially the claims by Smart Energy GB in their campaign for a smarter Britain. They state on their website, that Smart Energy GB is the campaign for a smarter Britain. They claim they are independent of Government, not an energy supplier and they do not fit Smart Meters. Their task is to support the rollout of Smart Meters so that everyone has the opportunity to share in the benefits that they bring for households, the economy and the environment! They say there will be no more estimated electricity bills, and the claim to save electricity consumption is vacuous. There is the misleading claim that a Smart Meter will allow customers can view their usage and cost at any time - this is dishonest as a customer needs a *separate monitoring device* to carry out this function as has been explained. It is a *sham* to claim a Smart Meter is good for the environment any more than a conventional digital or analogue meter - all types of meter simply measure power usage and cannot discriminate between the types of electrical generation that is being fed to a property. Why are the privatised power companies imploring people to have a Smart Meter fitted, telling the customer that it will help in using less of their product, which is electricity - it does not make sense - private companies are there to sell their product and make a profit for their shareholders! It is not surprising Robert Cheesewright and his chums at Smart Energy GB are keeping quiet on the £11 billion, and rising, expenditure for fitting Smart Meters. They are certainly keeping their heads down on the **coming compatibility nightmare, when 4G and 5G technology replaces 2G and 3G technology**. Perhaps the term Smart Meter is an appropriate name after all, as the hoodwinking and fabrication will surely make **a sane persons eyes *smart*!**

A Totem of UK Energy stupidity

Build Your House on the Rock

24 "Everyone then who hears these words of mine and does them will be like a wise man who built his house on the rock. 25 And the rain fell, and the floods came, and the winds blew and beat on that house, but it did not fall, because it had been founded on the rock. 26 And everyone who hears these words of mine and does not do them will be like a foolish man who built his house on the sand. 27 And the rain fell, and the floods came, and the winds blew and beat against that house, and it fell, and great was the fall of it." **Matthew 7:24-27 (English Standard Version)**

CHAPTER FOUR

'Justice will overtake fabricators of lies and false witness'

Heraclitus, Greek Philosopher (535 BC – 475 BC)

WIND ENERGY

Attempting to generate LARGE amounts of electrical energy from the wind in the UK is flawed and silly - attempting to rely on an energy source that is unreliable and variable is akin to building a house on sand.

Carpeting the British countryside and coastline with scenery destroying, bird and bat killing wind driven generators (they are not turbines – see later) surely has to be an act of not only wanton vandalism, but engineering ignorance and madness. Those who understand the technology and engineering fully know the policy will bring about a predictable energy crisis - brownouts and blackouts will become a fact of life - this is not scare mongering but simply due to the machines total dependence on the wind – remember the windless summer of 2021 which meant the wind farms produced less electricity, so conventional power plants had to burn more fuel than normal. Everyone knows the wind is variable and unpredictable, which will obviously result in an unpredictable means of generating electrical energy - but it would seem that Government does not want to acknowledge this fundamental truth – a quick scan of the G.B. National Grid Status website, (www.gridwatch.co.uk) will show the variability of wind generated electricity. The site not only shows the various sources of generation in *real* time, but offers the chance to compare wind generation from that of a windy day, to when there is very little wind. It will soon become very clear that placing a heavy dependence on this type of variable and unpredictable generation is an act of stupidity.

There appears to be no limit to the folly of National or Local Government in pursuing large scale wind generation in the UK, and it is the height of irony that nearly 2 million trees have been felled in Wales to accommodate wind farms. Such action clearly show the vacuous claim of the Welsh

Assembly of caring for the environment - chopping down trees cannot be defined as being environmentally friendly and demonstrates double standards. This is not to overlook the tons and tons of concrete used in creating generator bases and the necessary infrastructure to connect to the Grid. Huge areas of peat and gorse land have disappeared, land which would normally soak up rainfall, minimising run-off and the threat of flooding. Thus apart from the now greater risk to future flooding there will, under windless conditions, be the lack of electricity generation coupled with the loss of millions of environmentally enhancing trees – surely such planning and implementation is worthy of the Mad Hatter. It begs the question whether tree felling played a part in the flooding in south Wales and elsewhere during 2020 - remember trees soak up water - a case of chickens coming home to roost maybe?

National Resources Wales (NRW) have stated the number of trees that have been felled for all onshore wind farm development in Wales (at the time of writing) are as follows, Clocaenog Wind Farm: 307,200 Cefn Croes Wind Farm: 568,000 Pen y Cymoedd and Maerdy Wind Farm: 732,320 Brechfa Wind Farm: 330,880; the total number of trees felled for onshore wind farm developments on NRW managed land so far totals 1,938,400. A similar 'wind farm' environmental threat applies to Scotland and its wonderful wilderness and irreplaceable scenery - the magic of the Highlands will be truly lost - it will be a tragedy beyond measure.

Regarding smaller scale wind generation and for what it is worth, chatting recently to a electricity meter reader he offered that not one wind farm or single generator he had visited, have not had a problem with their wind generator installation - it was a pity he did not have the time to elaborate further. But this revelation should not come as a surprise as wind generators are mechanical devices and, apart from unforeseen breakages, accidental fire and lightning damage, these machines will obviously suffer from wear and tear, especially those being exposed to extreme weather situations at sea. Wind farms placed in the marine environment are the most exposed to the weather having to contend with corrosive salt air, high seas, gales and violent storms. This also begs the question of how the necessary maintenance or repairs will be carried out when the sea is too rough for landing a maintenance crew – such conditions place huge amounts of stress on both people and machines - landing maintenance engineers on generators at sea can prove very dangerous and thus require a high level of health and safety practices. My wife and I have witnessed a maintenance boat having to return to shore off the coast near Great Yarmouth during late August, due to a large swell making it difficult for the crew to land safely. The North Sea can of course be much more unforgiving during the winter months when breakdown is much more likely with conditions inhibiting attendance by maintenance engineers.

It is accepted there are places around the globe where the employment of small scale wind generators have an arguable case. These are at places such as cattle farms in the Australian out-back, places which make any connection to an electrical grid system practically impossible due to distance and economics; the same criteria applying to cattle ranches in the USA and other such similar areas. Obviously, when there is no wind or indeed too strong a wind, power cannot be generated, so it makes sense to have a backup source of electrical energy in the form of a petrol or diesel driven generator. Parts of the United Kingdom that could argue a case for a wind generator would be remote dwellings on islands off the coast of Scotland, where connection to the local distribution network or Grid would not be viable due to geographical placement and economics. But of course this need would have to be weighed in the balance with local environmental acceptance and threat to wildlife; some form of backup again would be required for a guaranteed power supply. The author recognises that at locations on the planet where there is a steady wind source, and there is no measurable impact on the local environment or wild life, then it would be churlish to deny large scale wind generation in such areas - but where are there such areas - this is far from the case for the United Kingdom and for many other parts of the globe. At 1305 GMT on 17th November, 2020, a very windy day across the UK the total demand for electricity was **34.89 GW**, of which renewable generation contributed 15.692 GW (45%), and the largest contribution was from wind at 11.431 GW (33%) of total demand. Solar was a derisible 0.922 GW (3%) of total demand, with 8.184 GW (23%) from gas. Then at 1325 GMT on 9th October, 2021, a not so windy day, the total demand was **30.04 GW**, with the lion's share from gas at 13.91 GW (46%), solar was 3.85 GW (13%) and wind only 2.616 GW (9%). Both wind and solar produced only 6.466 GW (21%) thus not a very impressive figure and calamitous if the UK were to be dependent on this form of generation.

A large wind generator requires a 'cut-in' wind speed of about 16 kilometres per hour (10 miles per hour) to start generating electricity, and they reach peak power output at around 53 kilometres per hour (33 miles per hour). At very high winds speeds, greater than 80 kilometres per hour (50 miles per hour), the wind generator has to be SHUT DOWN, otherwise structural damage will occur.

An electrical generator of any kind works on the principle that a conductor moving through a magnetic field has a current induced into the conductor. The conductor must be moving at a certain speed through the magnetic field to reach a certain current level, which when applied to the load, results in a certain output voltage. It is important to note the speed of an AC wind generator determines the frequency of the generated alternating current. It therefore follows that a wind generator must get up to a certain speed to generate a certain voltage and frequency of operation. At very low wind speeds, there is insufficient torque exerted by the wind on the wind generator blades to make them rotate. However, as the wind increases, the blades will begin to rotate, turning (propelling) the drive shaft to the gear box, and then a generator shaft to the generator to produce electrical energy. Therefore do not be deceived dear reader by slow moving generator blades, as the wind speeds at which the wind generator blades are able *generate power* for the network, is 16 kilometres per hour (10 miles per hour).

Before proceeding further it is useful to understand the basic workings of a wind generator, which can be summarised as follows:

- The large rotary blades (propeller) attached to the tower of a wind generator simply catch the wind and in doing so, employ this wind energy to turn a low-speed shaft.

- This low-speed shaft then, via a gearing system, turns a high-speed shaft which rotates at speeds between 1000 – 1800 revolutions per minute (rpm).

- The high-speed shaft connects to a generator to produce electrical energy.

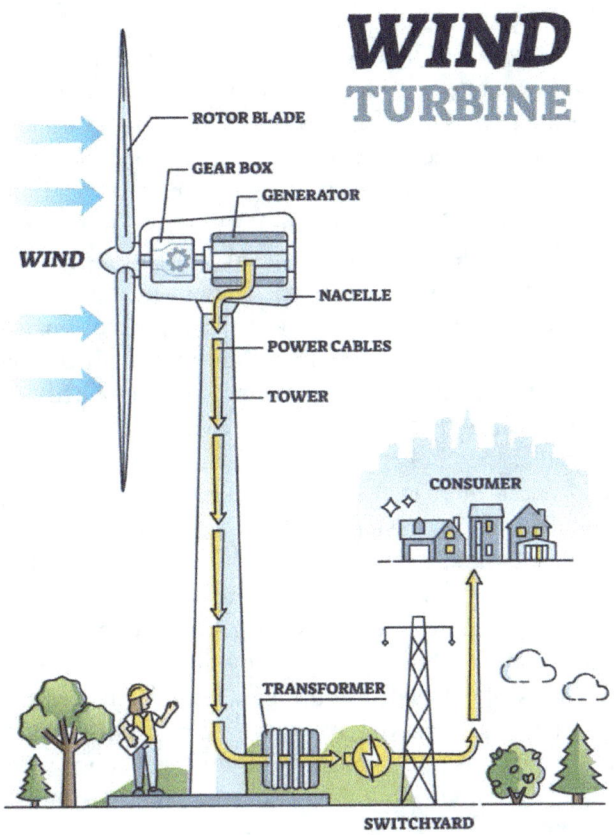

WIND TURBINE

ROTOR BLADE

GEAR BOX

GENERATOR

WIND

NACELLE

POWER CABLES

TOWER

CONSUMER

TRANSFORMER

SWITCHYARD

Example to the misuse of the engineering term TURBINE, when it should simply be WIND GENERATOR.

It is imperative for the reader to understand that when the generator blades are spinning (start-up speed); it does not mean that the generator is producing *meaningful* power. The start-up speed is the minimum wind speed needed for the blades and consequently the propeller shaft to begin spinning. At low wind and rotational speeds the wind generator will produce no power until the wind speeds reach the required cut-in speed for a specific wind generator. All wind generators have a distinct start-up speed and a cut-in speed. As the wind speed rises above the cut-in speed, and hence the rotational speed of the blades, the level of electrical output power rises rapidly. However, at around 53 kilometres per hour (33 miles per hour) the power output reaches the limit that the electrical generator is capable of. This limit to the generator output is called the rated power

53

output and the wind speed at which it is reached is called the rated output wind speed. At higher wind speeds, the design of the wind generator is such to limit the power to this maximum level and there is no further rise in the output power. How this is done varies from design to design but typically with large wind generators, it is achieved by adjusting the blade angles so as to maintain the power at a constant level. As the speed increases above the rate output wind speed, the forces on the wind generator structure continue to build and, at some point, there will be a risk of damage to the propeller blades. Thus a braking system is employed to bring the blades to a standstill. This is called the cut-out speed and is usually greater than 80 kilometres per hour (50 miles per hour). The cut-in and cut-out speeds are the operating limits of the wind generator. By keeping within this range, the available energy (useful power) is above the minimum threshold and the structural integrity of the wind generator is maintained. It cannot be reiterated enough times that at low wind speeds of less than 16 kilometres per hour (10 miles per hour) a wind generator does NOT produce any meaningful power, as there is insufficient energy in the wind to productively turn the propeller blades faster. Additionally, absolutely NO electrical energy is produced when the wind speed is greater than 80 kilometres per hour (50 miles per hour) as the wind generator is then shut down to prevent damage occurring. Therefore it makes for good common sense to carry out an assessment of the potential wind energy at a proposed site. No one in their right mind would install a wind generator in the Doldrums for the obvious reason to the lack of any wind, apart from any other factors. Conversely, it would be silly to install a wind generator where there are continuous and extremely high winds as wind generators, would be continuously shut down.

Large wind generators

Generating electricity from the wind is not a new concept as the technology goes back to the winter of 1887-88 when the American, Charles F. Brush (1849-1929) built the first automatically operational wind generator, in Cleveland, Ohio, USA; his generator was known as The Giant Brush Windmill and utilised a 12 kW direct current generator. It was a very large construction consisting of a large disc shaped rotor which had a diameter of 17 metres (50 feet), which incorporated 144 rotor blades made of cedar wood. The modern wind driven machines the Wind Industry employ are obviously of better design using the latest technology, materials and computer control techniques. But, remember, they are basically no more than an electrical generator driven by a shaft connected to a propeller, which is driven by the wind; the larger wind generators necessitating a clutch and gearbox between the propeller shaft and the generator. The author regards the term Wind Turbine as a clever marketing ploy by the wind industry, and to exacerbate the situation politicians and indeed the media ignorantly persist in the use of this misleading term. The description conjures up a much more sophisticated and powerful machine than it actually is. These wind dependent and limited machines are simply wind driven generators; no more, no less. A turbine is a fairly complex piece of machinery consisting of numerous blades along a shaft fitted *within a casing* i.e. an engine in which steam, water or gas is made to spin a rotating shaft by pushing on angled blades, like a fan. Turbines are among the most powerful machines with steam turbines driving generators in power stations and ship's propellers, with water turbines spinning the generators in hydroelectric power stations, and gas turbines powering jet aircraft. Charles Algernon Parsons (1854-1931) the inventor of the compound steam turbine must be turning in his grave (please excuse the pun). The steam turbine is a form of heat engine that derives much of its improvement in thermodynamic efficiency from the use of multiple stages in the expansion of the steam. Because the turbine generates rotary motion, it is ideal in driving the shaft of an electrical generator. Indeed, about 85 per cent of all electricity generation in the United States in the year 2014 was produced by steam turbines. Simply stated a steam turbine is a device that extracts thermal energy from pressurized steam and uses it to do mechanical work on a rotating output shaft. The wind driven machines the Wind Industry employ are not turbines, but simply, electrical generators connected directly, or via a gear box and clutch, to a large wind driven propeller. Simply have a look at one of these 'monsters' and judge for yourself. If that does not convince you dear reader, then visit your local conventional power station and ask the engineers if it is possible to view the steam turbines, or at least to describe and explain the workings to you.

The three bladed versions can be compared to a large scaled up version of a Lancaster bomber propeller. The propeller is an accurate description as it does indeed propel or drive the shaft (rotor) which drives the electrical generator, such as in the same sense as the propelling shaft between the gearbox and the rear axle in a motor vehicle. Therefore these propeller blades are not a turbine in the same context that the sails or vanes of a conventional windmill are obviously not a turbine. According to the wind industry logic should we now be referring to the large rotor blades of helicopters as turbines, or indeed any form of aerofoil? The truth is further laid bare if we look at turboprop aircraft which employ a compressor and turbine to drive a propeller. There are two main parts to a turboprop propulsion system, the core engine and the propeller. The core is very similar to a basic turbojet except that instead of expanding all the hot exhaust through the nozzle to produce thrust, most of the energy of the exhaust is used to turn the turbine. The drive shaft is connected via a reduction gear box to a propeller that produces most of the thrust. The exhaust velocity of a turboprop is low and contributes little thrust because most of the energy of the core exhaust has gone into turning the drive shaft. Thus a turboprop engine can be said to be simply a jet engine attached to a propeller; the turbine at the back is turned by the hot gases, and this turns a shaft via a reduction gear to turn the propeller: a number of small airliners and transport aircraft are powered by turboprop engines. Modern turboprop engines are equipped with propellers that have a smaller diameter but a larger number of blades for efficient operation at much higher flight speeds; to accommodate the higher flight speeds the blades are scimitar-shaped with swept-back leading edges at the blade tips.

When people first harnessed the wind to drive a mill they called the whole device a windmill (wind driven mill); when the wind was used to pump water the device was called a wind pump (wind driven pump) - are the sails of wind driven boats now to be referred to as turbines and are the sails (vanes) of a conventional windmill now to be called turbines – of course not, it is complete nonsense! Surely convention demands that using the wind to drive an alternating current generator, then it simply should be called a wind driven generator - the use of the word turbine is totally misleading. Common sense alone demands the most accurate term is a WIND DRIVEN GENERATOR (WDG), or WIND GENERATOR (WG)
for short, it is as simple as that for there is no ambiguity, or any pretension to other than what it is.

Wind farms

Again another misleading and irritating use of words is the term WIND FARM as this conjures up something ecologically friendly and productive, and therefore by association, the term wind farm implies a site that is very 'green' and productive. Supporters of wind farms will no doubt claim the wording is correct as the machines are indeed 'farming the wind' and the wording is correct. But unfortunately whilst some farms are very productive, and care for the environment, others are not. A wind farm is not a very productive farm being at the mercy of the wind and thus its output is unreliable and limited. The construction and installation of a wind generator produces significant amounts of CO_2 coupled with having a very significant *industrial* impact on the surrounding landscape – not very farm like.

A wind farm is little more than a large grouping of ineffective and ugly wind generators contained in a single site. The 'environmentally friendly tag' associated with wind generators which the Wind Industry would have us believe is greatly exaggerated. According to British Nuclear Fuels (BNFL) a wind farm covering an area of 712.25 square kilometres (275 square miles) would be required to produce the equivalent electrical output of Wylfa nuclear power station (now decommissioned). This nuclear station had an installed capacity of 980 MW and a footprint of less than 150 acres. Remember 1 square mile is equal to 640 acres, so we are comparing the *industrialisation* of 176,000 acres to a mere 150 acres. Thus it is not unreasonable to consider a wind farm of this size would be an act of mindless folly. It is vital to note the average electricity output of this 'huge' wind farm would merely match the output of the nuclear power station - that is when the wind is blowing at the correct speed. Additionally, it would not result in the closure of any fossil-fuelled or nuclear plant as unpredictable wind power has to be 'shadowed' at all times by a secure, controllable source of electricity such as fossil fuelled or nuclear powered stations. So much for cutting down on CO_2 emissions or reducing nuclear energy, a bit of a fraud, I think you will agree.

Backup

Due to the very nature of the wind in the UK a wind generator cannot be relied upon to produce an uninterrupted supply of electrical energy - under conditions such as high pressure in winter or summer, a wind generator or indeed a wind farm, may not produce any meaningful power at all - then at other times the vagaries of the wind have to be considered as at any one moment the wind may be blowing at a useful speed, then the next moment, just a breeze. Therefore it follows that some form of backup that can react to all these conditions will be required if security of supply is not to be in danger. The difficulty with backup for wind farms is the question of how much backup is to be kept running at any one moment in time. Running a continuous full backup by a conventional power station would be rather silly, as then there would be no need for the power from the wind farm and as such it would be wasted. So the question is how much 'spinning reserve' would be required that would be economically viable to satisfy a variable demand for power? Without going into too much detail spinning reserve is generation capacity that is on-line but unloaded (not used) and that can respond to compensate for generation or transmission outages within 10 minutes. A 200 MW power plant running at say 90 MW may be considered part of the spinning reserve, the spinning reserve in this example being 90 MW which is known as the reserve capacity. Non- spinning reserve (supplement reserve) refers to power sources which are presently not on-line, but can be switched into the system if the demand rises. The non-spinning reserve should be able available to supply the load within 10 minutes. The tricky part for the power producing industry is how to determine the total amount of backup for wind generation that is considered economically viable. Who can say for sure, where and when the wind will blow, and obviously the greater the number of wind farms, spread over a greater area, the bigger the problem.

If a single wind farm is connected intermittently to the National Grid it would not pose too much of a problem (just synchronisation) as the Grid has approximately a 20 per cent margin over peak demand, this being required to guard against line transmission or generator failure. The real problems start to arise when *more and more* onshore and offshore wind farms are connected to the Grid as apart from synchronisation, there will come a point when serious consideration has to be given to providing additional resources, with the ability of being able to assess correctly the capacity and number of stations necessary to ensure system security of supply.

ONSHORE WIND FARM

It is important to realise that you cannot simply switch on steam driven turbines similar to the action of throwing a light switch and have immediate power. It is necessary to activate a heat source at the power station to convert water into steam to drive the turbines – this takes time. Coal-fired stations will take several hours from cold before they can produce any useful energy. Gas, oil and nuclear stations still have to boil the water for steam, although they can be brought on-line quicker than coal.

One of the most bizarre arguments put forward by proponents of wind machines in the UK is the claimed *significant* saving in carbon emissions. How can this be, when as we have seen, wind generators for a guaranteed continuance of supply, require backup by conventional power stations. The vacuous argument of UK wind power supporters is put into even more context when the pollution from the industrialisation of China, India and Brazil are brought into the equation, noting that China is the planet's top polluter! It truly is a vacuous argument considering the UK contributes to **less than 1 per cent of global emissions.** Do the supporters (Greens) of wind power not realise that during 2005 China set out on a seven-year programme to build over 500 new coal-fired power stations, do they not know that most of China's electricity is generated by burning dirty coal in dated power stations creating horrendous air pollution. Nearly half a million Chinese are killed by smog each year, perhaps the Greens are not aware that it is predicted that in approximately ten years' time there will be 140 million vehicles on China's roads. As a comparison, there were just fewer than 40 million registered vehicles in the UK during 2019. Are the Greens also not aware of the massive building programme going on in China demanding millions of tons of concrete and other associated materials - imagine the impact on global CO_2 emissions!

OFFSHORE WIND FARM

Then there are the high atmospheric emissions from America, India, and Brazil - why are they isolating and picking on the UK? Pro-wind farm supporters need to *get real* and put things into perspective - no doubt they will go into denial when confronted with the truth, and attempt to justify their ludicrous actions by arrogantly claiming that the UK must set a global example. It is important though to discriminate between CO_2 emissions and pollution such as carbon monoxide and very small particulates. Indeed, CO_2 is necessary for all life on Earth and a reasonable additional amount to the atmosphere can be beneficial and will manifest itself by increased plant growth. That is why the Dutch pump CO_2 into their greenhouses – not to poison the plants inside, but to enhance healthy growth. But of course, just like other things in life, too much can be a problem.

We are all in favour of a clean and pollution free environment and accept there needs to be an intelligent and balanced approach, and it will do well for wind farm supporters to question the cost of not only the construction of, and the running of a wind farm, but also the provisioning of a site with its concrete foundations, access roads, transformer buildings and wires/cables/poles for connecting to the electricity network. The construction and erecting of wind generators require thousands of tons of aggregate and concrete for their foundations, not to mention the material for the approach roads required for the ongoing repair and maintenance of the wind generator. Materials such as cement consume large quantities of carbon-based fuels, such as coal, gas or oil; emitting large quantities of greenhouse gases when burnt in cement kilns during the manufacturing stage. Additionally the aggregate required in concrete making also demands carbon-based fuels for the machinery used in extraction/quarrying process. This is not to overlook the necessary transportation of the material. All these procedures contribute to the considerable amounts of atmospheric emissions such as carbon dioxide; there is the erection of unsightly poles and/or possibly pylons for connecting to the local distribution or grid network. Then there is the inexcusable industrialisation and desecration of the local countryside to consider with its consequential threat to tourism. For example, large numbers of wind generators across the Welsh hills will be devastating for the countryside and the people who live and earn their living there. Tourism is the largest rural industry, and earns over £2 billion a year for Wales, and as such it contributes 7 per cent to Welsh GDP, far outweighing agriculture, which contributes less than 2 per cent of GDP. It is by far the most important rural earner.

Wind generators create an unacceptable adverse effect to the character and visual qualities of any landscape. Indeed, allowing generator numbers to increase creates an adverse cumulative visual impact on the countryside. All Planning Offices in the UK should automatically dismiss any application for a wind farm or a single wind generator unless there are extremely good mitigating reasons to allow an application, and then, only with the agreement of the local populace. Welsh Government Ministers claim wind farms will minimise the carbon footprint of Wales. This is hypocrisy when they are allowing the felling of trees (the lungs of Wales) only to be replaced by an industrialised landscape with ineffective wind generators. You truly have to wonder what motivates these people, and I have yet to meet a politician that can offer an intelligent, reasoned and informed case for large scale wind generation in Wales, let alone the UK.

Keith Anderson, the Chief Executive of Scottish Power, said that during 2018 Scottish Power sold its last remaining fossil fuel power stations for £700 million, becoming the first energy business to rely only on wind farms to generate electricity. He also said the company planned to invest £1.7 billion in the UK spending £4 million a day on renewables across 40 wind farms. This is astounding for if Scotland is going to be totally reliant on wind farms, then what will be the situation under the condition of little or strong winds? I do not think the population north of the border are going to be overjoyed when their lights go out - but then perhaps he has not offered the whole story under such conditions - will electricity be imported to Scotland from the new Norwegian Interconnector, namely the North Sea Link (NSL) which has a capacity of 1,400 MW (see chapter on Alternative Energy) and the predominantly fossil-fuelled electricity of the National Grid south of the border? It was also unbelievable that the Chief Executive said England was not windy enough for wind farms – does the wind never blow, for example, across the Lake District, Yorkshire Moors, the Pennines, or Dartmoor - but it was OK to build wind farms in Wales (thank you Mr Anderson for offering to desecrate another beautiful country) and I naturally assume that Mr Anderson thinks the wind stops at Offa's Dyke! It is revealing that Mr Anderson studied accountancy at Napier University, worked for Standard Life and the Royal Bank of Scotland (RBS) as an auditor before joining Scottish Power in 2000. So, with the greatest respect, it would appear Mr Anderson is not highly schooled in engineering, technological or environmental skills, but simply in the practice of recording, classifying, and reporting on financial transactions for a business - which figures - sorry about the pun. In his *energy assessment equation*, which should encompass a number of *variables,* apart from financial matters, it does seem that variables such as reliable generation, important environmental issues and wild life, have been omitted. Recognise also that Scottish wind farms are paid millions when they are forced to shut down due to high winds which obviously results in no electricity being generated – something attractive for an accountant – being paid something for nothing. It would appear the Chief Executive has no feelings or concern for the irreplaceable beauty of the

Highlands, when wishing to desecrate Scotland with hundreds and hundreds of monstrous industrial size wind generators, compounding this felony with the slaughtering of numerous wild birds and bats, which it appears are not going to keep him awake at night!

The mindless slaughter of wild birds is a violation which thankfully, Save the Eagles International (STEI) and the World Council for Nature (WCFN) have a lot to say about. Such that it is contrary to the truth, to pretend that these industrial structures are *carefully sited*, so as to avoid risks to birds and bats. It is equally false to allege that grouse and other ground-nesting birds do not mind laying their eggs under wind generators, or that raptors avoid these dangerous areas. Mark Duchamp, President of STEI (www.savetheeaglesinternational.org), and WCFN has noted that many wind farms have been placed in the worst possible locations, where they will mince Scottish eagles into extinction: Eishken (aka Eisgein or Eisgen), Pairc, Pentland Road, Edinbane, Ben Aketil, various eagle ranges in Argyll, et cetera. 'Hypocrisy and deceit are rampant' laments Mark Duchamp. The U.S. Fish and Wildlife Service and American Bird Conservancy say wind generators kill 440,000 bald and golden eagles, hawks, falcons, owls, cranes, egrets, geese and other birds every year in the United States, along with countless insect-eating bats. Although new studies reveal that these appalling estimates are frightfully low and based on misleading or even fraudulent data. The reality is that in the United States alone so-called eco-friendly wind generators kill an estimated 13 million to 39 million birds and bats every year. To put it in a nutshell, THERE WILL BE SO MUCH LOST, FOR SO LITTLE GAINED. It can
be described as the pursuance of a policy dreamt up by lunatics. It is disgraceful and to their shame, that in the 21ˢᵗ Century supposedly educated, intelligent and knowledgeable leaders have succumbed to the political panic of climate change, which has triggered a lemming style rush for renewable energy sources without an open and intelligent discussion of the merits and drawbacks of each.

I wonder how many eco-warriors are aware of the global mining that contributes to materials used in wind generators, especially the pollution this mining causes. As an example, neodymium is a *so-called* rare earth element, a silvery metal with a very important role in renewable energy. When combined with iron and boron, it makes strong magnets that are important both for generators in Wind Generators and for the electric motors in Electric Vehicles (EVs). It should be noted that rare earth elements like neodymium are not that particularly rare - it is a metallic chemical element classified in the rare earth group of the periodic table. Although the term 'rare' is used in its elemental group, neodymium is actually relatively abundant. About 85 per cent of the world's neodymium comes out of a few mines in China - about 30 per cent of the planets deposits are located there. One mine in Baotou, Inner Mongolia, China has created a toxic lake and other environmental horrors - this toxic lake was created by damming a river and flooding good farm land. Neodymium is referred to as rare earth material, but it is actually no rarer than copper or nickel and is relatively common across the planet. The problems arise in the extraction process, which are hugely hazardous and toxic, producing large amounts of poisonous waste as a by-product. Shamefully this industrial process highlights China's willingness to suffer environmental damage that other nations would not tolerate - out of sight, out of mind, is no excuse for the Western Nations. I very much doubt if most eco-warriors or indeed the general public know of Baotou, but the mines and factories there help to keep our modern lives ticking. A trip to the Baotou Steel Rare-Earth Hi–Tech Co. Ltd, the largest industrial city in Inner Mongolia will give many an eco-warrior something to think about and realise the 'green' picture they and politicians paint about wind and solar generation is an illusion. A study by Massachusetts Institute of Technology (MIT) in Cambridge, Massachusetts, estimates that a 2 MW wind generator contains about 752 pounds of rare earth minerals.

The excellent researched book, 'The Wind Farm Scam', by Dr John Etherington (sadly now deceased), argued as far back as 2009 that wind generators could not produce enough electrical energy to reduce global CO_2 levels to a meaningful degree. Wind power is by nature intermittent and cannot generate a steady output, necessitating backup coal and gas power plants that significantly negate (especially coal) the saving of greenhouse gas emissions. In addition to the inefficacy of wind power there are ecological drawbacks, including damage to habitats, wildlife and the far-from-insignificant aesthetic drawback of the assault upon natural beauty and the pristine landscape, which wind generators entail. Dr Etherington (rightfully) argued that wind power has been, and is being, excessively financed at the cost of consumers who have not been consulted, nor informed that this effective subsidy is being paid from their bills to support an industry that cannot be cost efficient or, ultimately, favour the cause it purports to support.

It was extraordinary that a British Prime Minister, namely Boris Johnson, has actually said (Sept, 2020) that he wanted to turn the UK into the 'Saudi Arabia' of wind power. Astonishingly he opined that "As Saudi Arabia is to oil, the UK is to wind - a place of almost limitless resource, but in the case of wind without the carbon emissions and without the damage to the environment." Additionally, Mr Johnson opined that we have got huge, huge gusts of wind going around the north of our country such as Scotland. Well! Mr Johnson is correct on one issue, and that there are huge gusts around Scotland, and it simply confirms how ill-informed he is – for it is this *very* **reason** why wind farms have to close down so often in Scotland - it is to save wind damage to the wind generators which obviously results in nil generation. It is scandalous that these wind farms when they are NOT producing any power are still being paid millions of pounds – this simply means that future power bills have only one direction and that is upwards.

Roman emperors used the tactic of 'bread and circuses' to keep Roman citizens happy. Mr Johnson the Latin Scholar, also appears to admire the grand gesture by Roman emperors – could this be the motivation behind the 'Saudi Arabia' grand dream and gesture? In reality it is the stuff of nightmares that Boris Johnson envisages the UK will derive most of its future electrical energy from wind farms. It clearly demonstrates Boris Johnson and his Ministers have minimal engineering or technical knowledge and are ignorant of how the power industry functions. For example, it is very telling that the Rt Hon Anne-Marie Trevelyan, Minister for Energy, Clean Growth and Climate Change, at the Department of Business, Energy and Industrial Strategy from January 2021 to September 2021, is a chartered accountant by trade. This obviously begs the question, and with the greatest respect, if the lady has any worthwhile and contributory electrical engineering knowledge or experience in the power industry, not to mention schooling in Earth sciences and climatology? I very much doubt, dear reader, if you would telephone a chartered accountant for help when your plumbing starts to leak or the lights go out
– everyone to their own expertise. These are serious issues as Ministers are responsible for UK energy security, keeping industry and business going, and the vital need to make the sensible decisions – if not, we will all suffer as a consequence of their failings!

To put Boris Johnson and his Ministers ignorance into context it would have meant that the 346 TWh annual consumption of electricity for 2019 would require the output of 33 large CCGT Power Stations, each of 2000 MW capacity working with an efficiency of 60 per cent. But Boris Johnson's craze for wind technology would mean dependence on a mind- boggling total of **76,000 onshore wind generators**, each having a capacity of 2 MW, and due to the nature of the technology involved are restricted to an efficiency of 26 per cent. If all the generators were sited at sea then that still would be **49,000** working at 40 per cent efficiency - where on earth (or sea) do you site 76,000 or 49,000 wind generators, remembering that fossil-fuelled power stations are reliable and controlled by humans, whereas wind generation is totally at the mercy of the elements. See Appendix Four for further detail.

Comparing the number of power stations and wind generators for the UK 2019 peak demand and annual demand we have:

	Load factor	Peak demand (47.275 GW)	Annual demand (346 TWh)
Power Station	60%	39	33
Onshore Wind generator	26%	90,913	76,000
Offshore Wind generator	40%	59,093	49,000

During the summer heatwave in 2018 it was reported that more than £1 Billion was wiped off the value the Scottish and Southern (SSE) Company, after wind generators came to a halt due to the warm, still weather. No doubt SSE shareholders are thrilled and wondering why SSE is investing in ineffective and scenery destroying wind farms - no doubt wondering how much electricity their wind farms will be generating in the future during winters stormy weather, especially those sited offshore. Amanda Holden, as reported in the Media during January, 2022 is supporting a campaign to halt an off-shore wind farm, whose generators will be taller than the Eiffel Tower. Yes, that is correct the height to the top of the upper blade from sea level will be 325 metres (1,066 feet). There are plans by the German energy company RWE to erect 116 wind generators eight miles off the coast of West Sussex, covering an area about the size of the Isle of Wight. It is claimed the wind farm will generate enough energy to power one million homes – really - for that to become a reality, will require the wind to blow continuously and at the right speed. As mentioned earlier wind generated electricity in the UK fell during 2021 due to lack of wind. A summer of low wind demonstrated how wind farms are ineffective and expensive - calm weather over Europe during summer 2021 helped drive peak hour power prices to their second highest level since 2018. Energy companies, including SSE in the UK and Orsted in Denmark, reported their lowest wind speeds for two decades. Although I should point out that Ms Holden has joined Tory MPs, environmentalists and locals in urging RWE to locate the wind farm further out to sea or in the North Sea, so that it is not visible from the land. But this is short- sighted (excuse the pun) as this proposed wind farm should be completely abandoned, and not just a case of out of sight, out of mind. It is important to fully realise that in high winds, generators must be halted because they are easily damaged. Additionally, the build-up of dead bugs has also shown to significantly reduce the maximum power generated by a wind generator - reducing the average power generated by 25 per cent and more. But adding more to the misery for off-shore generators, the build-up of salt on the blades similarly has been shown to reduce the power generated by 20 per cent to 30 per cent.

The reader should now have a better insight as to why their energy bill is increasing – ever going up – electricity from the wind does not come cheap, coupled with the undeniable fact that it is variable and unreliable, and thus threatens UK energy security. I would wish to finish this chapter by including a letter I sent to the press at the start of 2022 which sums up the complete nonsense of large scale wind generation in the UK.

Dear Editor

Calling a spade a spade

The wind industry is truly a house built on sand – the sand in this case being human gullibility. It is this human failing that must have inspired and motivated Hans Christian Anderson to write the folktale of the Emperor's New Clothes. Every time that Government, politicians, the media and the public use the misleading term Wind Turbine, the wind industry must surely tip its hat and chuckle to itself at the widespread gullible acceptability. How many readers appreciate that the monstrosities employed by the wind industry could equally be called gas turbines, as the word 'gas' is interchangeable with the word 'wind' which describes the natural movement of atmospheric gas (oxygen, carbon dioxide etc) relative to the Earth's surface. Additionally, the employment of the word 'turbine' is completely misleading, and Charles Algernon Parsons, the inventor of the steam turbine, must be spinning in his grave, please excuse the pun. I would concede though that the term wind turbine sounds much more 'sexy' than, say, plain old 'wind driven generator'. But the term wind turbine falsely conveys an image of an efficient machine, as 'genuine' turbines are very efficient. Indeed, every fossil-fuelled and nuclear power station makes use of efficient gas turbines to turn their generators to produce electricity. It should be noted that the gas employed in these power stations is 'steam', remembering that steam is the name given to the gaseous state of water at or above 100^0C. Hydro-power stations are the exception, which use water to turn their turbines. The complete nonsense of large scale wind generation is laid bare when reliance is placed on these environmentally challenging monstrosities, which are totally dependent on the unpredictable and variable nature of the wind. These medieval machines, although employing modern technology in control mechanisms, are simply a high mounted generator connected, via a gear box, to a large two or three bladed propeller. We do not refer to the large blades of a helicopter as turbines and certain aircraft utilise a turboprop engine, which is a turbine engine that drives an aircraft propeller. No doubt, if the wind industry had its way, Boris the man who would turn the country into the Saudi Arabia of wind energy, would be crowned Emperor – with only a small boy seeing the man for what he truly is! Make no mistake dear reader there is a storm coming (power cuts etc) coupled with eye watering hikes in energy bills, and then everyone will be cursing the 'beast' that is called wind turbine!

End of letter

Having read so far, no doubt you will be appalled at the deceptions, misinformation and falsehoods. But please read on and be astounded at the ridiculous pursuit by Government to the madness of large scale solar generation in the UK. The short-sightedness of forcing electric vehicles (EVs) on the nation, without due thought to the infrastructure, and where indeed the extra electricity will be coming from; not to mention replacing the nations conventional boilers with heat pumps. Disconcertingly the following chapters reveal further chicanery such as the unintelligent and myopic acceptance of nuclear fission power stations and the deception and exploitation by the privatised water companies. But all is not doom and gloom as the chapter explaining Alternative Energy will demonstrate. Fracking and the employment of efficient gas-fired power stations can offer energy security and cheaper bills - leading to the potential of energy from space and nuclear fusion. It just necessitates the will and drive from an intelligent and knowledgeable Government.

CHAPTER FIVE

'A place for everything, and everything in its place'

Benjamin Franklin (1706-1790)

SOLAR ENERGY

It is not surprising that during the winter months UK solar generation will be extremely low. Indeed my 4 kW capacity roof-mounted 16 solar panel array will struggle to produce more than a few hundred watts of electricity on a cloudy January day. If this situation is scaled up to a field or more of solar panels, it is not difficult to appreciate that with the capital cost for construction, loss of land and the industrialisation of the countryside, it is madness to consider **Large Scale Solar** electricity in the UK - it will be utterly disastrous if the National Grid had to rely on this source of power generation. It should be understood that predominantly, UK electricity generation comes from gas - with unpredictable and variable amounts from solar and wind, which at night and during a large anti-cyclone, offer miniscule generation. The nonsense of large scale solar is compounded by the absolute madness of Scottish wind farms being paid millions, as a result of having to close down when the wind is too strong. Surely only the Mad Hatter could have dreamt up such a generation scheme where wind farm companies are paid for not producing any electricity! In assessing the merits of solar panels for producing significant amounts of electricity, it is important to recognise the latitude and weather patterns across the UK are not ideal for large scale solar generation. It is common knowledge how cloudy and gloomy the winter months can be, not to mention the short days, especially in Scotland. The British summer can also be very disappointing when the Sun refuses to shine, and we are all aware that the Sun does not shine at night. It would not be surprising if a misguided politician foolishly thought solar panels might function at night, arguing that moonlight is reflected sunlight. Ask your local politician if they support large scale solar parks (and indeed wind farms) in the UK, as their answer will tell you a lot about their engineering, technical and climate/weather learning, revealing their grasp of the *real* world around them. It is unbelievable that Government allows and encourages the widespread growth of large solar parks over the beautiful British landscape. It is total madness with acres and acres of land being covered by solar panels. Are our decision makers so stupid not to recognise that solar farms will obviously need backup from other sources of generation – it begs the question as to who advises Government and what their motivations are?

At the time of writing there are plans (May 2021) to build a very large solar farm at Langford, near Cullompton, Devon, which would see 91,000 solar panels built across 17 different fields, with the site spanning 61 hectares and surrounded by a four-mile long fence. The massive solar farm in Mid Devon will be bigger than the Vatican City. It is claimed the £40 million development would have an export capacity of 49.9 MW, and would be able to meet the energy needs of approximately 10,077 homes in Mid Devon. Really! What happens, for example, at night, or on a very cloudy and cold winter's day? The firm behind this enterprise claims that large-scale solar farms are needed to reduce emissions in the UK - but to almost paraphrase Mandy Rice-Davies, 'They would say that, wouldn't they'. The firm is talking utter nonsense as in reality large scale solar farms in the UK will have as much impact on global emissions as trying to save the Titanic with a stirrup pump. Nevertheless, expensive mains electricity, experience and common sense have shown solar technology can be deemed viable for **Small Scale** generation when fitted to buildings in the UK. My experience has shown that a relatively small array, such as a 4 kW roof mounted Photovoltaic (PV) system, will generate, on average, 4000 kWh per annum in west Wales - that is, 4000 units less of electricity needed from conventional power stations each year. It should be noted the average three-bedroomed house will use about 3500 kWh per annum, and solar generation can occur when there is no demand for household electricity, such as when the family are spending the day at the seaside, or in the country. This is when the solar generated electricity is diverted (exported) to the local electricity network. More importantly it is not a critical situation when the Sun is not shining, such as at night, as the consumer can fall back on mains electricity. Bearing in mind the *continuing rise in energy charges*, small scale solar arrays are an attractive proposition, even without government subsidy, as we shall see.

Roof-fitted solar panels

Wherever possible, it is recommended that every suitable roof should have solar photovoltaic cells attached to them, or *imbedded* in the roof at the time of construction. Roof-fitted photovoltaic cells help in diminishing the need for unfriendly environmental means of generating electricity such as coal. It is important though to recognise that roof-fitted solar arrays are *totally dependent* on a mains supply to enable solar generated electricity to be fed to a house. If the mains electricity fails, for whatever reason, the solar array inverter will automatically *disconnect* the solar panels from the incoming mains connection. The reason for tripping out the solar array is safety - the 'isolation' of any solar generated electricity is to stop this power entering the local electricity network where engineers could be working on the system. A similar isolation technique will be employed for a home connected wind generator, and unfortunately during a BLACKOUT, wind generation (or solar), will be TOTALLY INEFFECTIVE for providing power to the home. On Tuesday 1st June, 2021 our house in Ceredigion suffered from a blackout, followed by a brownout (voltage level 171 volts), then another blackout with the whole episode lasting for about 3 hours, during which the solar array was 'isolated' (safe). The mains failure problem was said to be a power wire down on an 11 kV route in the local network - our house is situated in a rural environment. It is somewhat ironic that home fitted so-called environmentally-friendly solar panels (or a wind generator) are, in a sense, totally dependent on predominantly fossil-fuelled power station generation. But it should be noted that if any dwelling is off-mains and self-sufficient, the situation will obviously be very different. The question has to be asked as to why the Government is not offering much more attractive inducements to have solar photovoltaic cells fitted to properties, indeed why is it not compulsory for new buildings to have **Flexible Thin Film Solar Photovoltaic Cells** imbedded in their roofs – see later in this chapter. These steps will benefit us all, and perhaps after digesting this chapter, readers will lobby their MPs for a return of an attractive **Feed-in Tariff**?

Feed-in Tariff (FIT)

The Feed-in Tariff scheme offered cash payments to households that produced their own electricity using renewable technologies, such as solar PV panels. Unfortunately at the end of March, 2019 the scheme closed to new applicants, but this did not affect existing instalments that had qualified for the payments. Indeed, the payments were guaranteed by the Government and paid for by a levy on everyone's energy bills. Those who signed up to the Feed-in Tariff when it first launched in 2010 were paid a much more generous rate than those who signed up shortly before the scheme closed. If readers are looking to install solar panels now, it will be necessary to qualify for the Smart Export Guarantee (SEG), which we shall examine in more detail later. Feed-in Tariff payments were made up of two elements, payable for up to 20 years (25 years if signed-up before August, 2012) and were paid each quarter. They were tax-free and Index

Linked from April each year.

There were three ways to earn and save money from the Solar PV Feed-in Tariff:

- **The Generation Tariff.** This paid for each unit (kWh) of electricity that was produced, regardless of whether or not it was used by the consumer. At the time of writing, for existing installations, the FIT paid 17.16 pence per unit of electricity generated.

- **Export Tariff.** For installations without an export meter the Export Tariff paid an additional fee. This was paid for electricity that was not used and was exported back to the National Grid; the amount of energy exported was deemed to be 50 per cent of that generated Thus if the system generated 4000 kWh over a period of time, then 2000 kWh was considered to have been exported over the same period of time. At the time of writing, an additional 5.50 pence is being paid for each unit (kWh) exported for existing installations.

- **Electricity Bills.** Substantial bill savings are to be gained if the electricity that is generated is consumed at source - a consequence of fewer units (kWh) on the power bill.

The Feed-in Tariff was very generous from 2011 to 2012, when the generation tariff paid was 54.17 pence for every kWh generated and 3.82 pence for every kWh deemed exported.

This meant a 4 kW system that was capable of generating 4000 kWh per annum would have been paid £2,243 a year and **£44,864 over 20 years.** This is not to overlook the electricity being saved by the household. Unfortunately the payments have significantly diminished since then, demonstrating that a Labour Government has been far more generous to the public than the Conservatives - surely it is putting salt into the wound when a Conservative Government abolished the Feed-in Tariff in 2019 and during 2020 the British Prime Minister, Boris Johnson announced his 'Saudi Arabia Wind' venture, which will cost the tax payer billions.

ROOF-MOUNTED SOLAR PANELS

Domestic solar generation

To illustrate the efficacy of domestic solar generation, two working systems are examined. The first being a four bedroomed detached house situated in the village of Boncath, Pembrokeshire, with solar generation over the period 1999-2013. The solar power array was installed on the roof of the detached garage at the property, with the panels having a tilt of approximately 35 degrees and a southerly orientation. The solar array consisted of ten 230 watt Dimplex high performance polycrystalline solar PV modules, offering a total system capacity of 2.3 kW. The capital cost of the installation was £7,818. Apart from the Feed-in Tariff payments the system reduced the annual electricity bill from an average of £500 per annum to £294.4 per annum, a reduction of over 40 per cent. Although it should be noted that some of the electricity savings were achieved by the employment of low energy lighting. The installation of the roof mounted solar array, also helped in achieving a reduction in the amount of Liquid Petroleum Gas (LPG) used for cooking and heating. The mains electrical power consumption before the installation of the solar panels was of the order of 5000 kWh per annum. During this period a mix of incandescent, CFL and LED bulbs was used for lighting – efforts were also made to purchase low energy rated white goods. To illustrate the effectiveness of roof-fitted solar panels it should be noted that the weather for 2012 was on the dismal side.

Pembrokeshire solar generation (kWh) for 2012

JAN	FEB	MAR	APR	MAY	JUN	JUL	AUG	SEP	OCT	NOV	DEC	TOTAL
42.5	93.1	219.4	181.8	306.9	241.3	257.1	230.4	219.6	137.7	83.2	43.2	2056.2

During 2012, rain and floods occurred during the early part of the year, followed by a sunny March and April - but this was followed by a poor summer. Generally the weather was quite dismal for 2012, which is reflected in the solar generation. It was telling that the solar generation for May was greater than June by 65.5 kWh where it would be expected the greater generation should occur during June, as this month, apart from longer days around the summer solstice is usually sunny with the strongest sunlight. Therefore 2012 was not considered an average year in the context of sunshine hours. But with everything considered it was concluded the solar array performed better than anticipated. As mentioned earlier, the average annual power consumption from the local electricity network from 1999 to 2013 for the property at Boncath was of the order of 5000 kWh per annum. The solar generation (2056.2 kWh) during 2012 turned out to be over 40 per cent of the mains consumption, and this being achieved during a predominantly cloudy year.

Solar energy earnings Pembrokeshire

During 2012 the solar generation FIT payment for the house in Pembrokeshire was a generous 43.3 pence per kWh generated. It was deemed, by the electricity company that 50 per cent of the total solar energy generated would be fed back (exported) into the local network and as such paid 3.1 pence per kWh exported. During 2012 a total of £922.2 was paid for solar energy generated, although it should be noted that for simplicity sake, the initial value of the FIT for calculations was used, as due to index linking (at April) the value of the FIT increased to 45.4 pence for each unit generated and 3.2 pence for each unit exported – using these figures would make the earnings marginally higher. It should be noted that not all the solar generated electricity was consumed in the house in Pembrokeshire, as a certain amount of the solar generated power had in fact been exported to the local power network. A good example being when export took place on a sunny day in the summer, when the occupants had spent the day at the coast, leaving the house empty - a minimal need for electrical energy such as low energy demanding items as fridges and freezers - the property in Pembrokeshire contained a fridge in the kitchen and two freezers in the garage. Bearing in mind that at other times of the year during the winter months, when solar generation is low, most, if not all solar generation was used at source. For simplicity sake the electricity company Standing Charge is not considered. Not having a separate 'export meter' it was difficult to know *exactly* how much solar generated electricity was exported, so it was assumed that all solar generation was consumed resulting in the mains electricity bill of £294.4 during 2012, instead of the average annual bill of £500. This gave a saving of £205.6 coupled with an addition of £922.2 being paid for solar generation under the FIT scheme. (See Appendix Three for further detail).

Both earnings from the FIT and mains power savings were combined and defined as INCOME. Therefore total income for the year was £1,127.8. Projecting this income over 20 years would amount to £22,556. (Note the saved mains electricity figure of £205.6 per year over 20 years would result in £4,112). Total income over 20 years was determined by deducting the initial capital cost of £7,818 for the solar PV system, giving a total of £14,738. This is deemed very acceptable both in energy conservation, savings, and as a sound investment – where else would you get such a return on your money. It should also be noted that with an income of £1,127.8 per annum the initial capital cost of the solar PV system is recoverable in just less than 7 years. The house in Pembrokeshire depended on gas (LPG) for cooking and central heating and apart from the savings in electrical energy during 2012 there was a noted reduction in gas usage as the house heating was complimented by the careful use of electrical heaters. Compared to current prices, the cost of the above solar PV system was relatively high – but the energy savings and the return on capital were proven to be very attractive due to the fact that Feed-in Tariff payments then were very generous. Today, although the FIT payments have now been replaced by the Smart Export Guarantee, system installation is that much cheaper, and the reader should now be able to have a 2.5 kW system installed for less than £4,000 – I have seen adverts for 4 kW capacity systems priced just under £5,000 - although I would caution about not buying too cheap as at the end of the day you get what you pay for. In the next solar panel example, a total of £7,100 was paid for a 4 kW system, almost twice the generating capacity as the house in Pembrokeshire. It was considered a good price considering the standard of workmanship and professional approach to the installation by a qualified local installer – also there is the security, assurance and peace of mind that a good local installer offers - being easily contacted should any problems arise. As we have seen, the Labour Government were very generous in the FIT payment, and although the FIT payment has been replaced by the less generous Smart Export Guarantee, the installation of a solar PV system is still considered a viable and attractive proposition from the saving in electricity and other energy costs; regardless of Tory Government miserliness it is prudent to remember that *energy costs are always rising* for when did the 'big six' energy companies ever offer a price reduction! Since the turn of the century, the average energy bill has doubled according to The Office for National Statistics. The next example involves a five bedroomed detached dwelling and considers not only a solar PV system but also a Solar iBoost unit that has the ability to identify and enable the consumption of what is defined as 'exported solar energy' at source.

Five bedroom detached property

During the month of May, 2013, after moving to a five bedroomed detached property on the outskirts of Cardigan, Ceredigion, and having been very pleased with the solar panel installation in Pembrokeshire, it was decided to install, during December, 2013, a total of 16 solar panels mounted on the roof of the detached property we had moved to. The roof of the property in Ceredigion has a tilt of approximately 40 degrees and a south westerly orientation. The solar array consists of sixteen Hyundai 250 Watt Black Frame Polycrystalline panels with a total system capacity of 4 kW at a cost of £7,100. The annual mains power consumption, without solar, was historically assessed at 6000 kWh per annum – this assessment being based on a five bedroom dwelling and the use of incandescent light bulbs. The house heating is by oil (under-floor heating on the ground floor and radiators upstairs) with an annual oil consumption of 1500 – 2500 litres depending on usage and how cold the winter proved to be. Cooking is by a combined electrical and gas (LPG) stove - with LPG gas consumption approximately 38 kg per annum. It should also be noted that some backup heating is provided from a wall fitted LPG gas fire in the lounge during extremely cold weather – which also offered an aesthetic effect to the lounge.

Ceredigion Solar Generation (kWh) for 2014

JAN	FEB	MAR	APR	MAY	JUN	JUL	AUG	SEP	OCT	NOV	DEC	TOTAL
88	197	331	424	471	658	567	486	424	219	166	100	4131

Similar to the Pembrokeshire calculations, the value of the FIT payments for Ceredigion at the start of generation in January, were used for the whole of the year and in this case amounted to 14.9 pence per unit (kWh) generated and 4.64 pence for each unit exported. Again it was deemed by the electricity supply company that 50 per cent of solar generation would be exported back into the local network. During 2014 a total of 4131 kWh was generated by the solar panels, the total FIT payment amounted to £711 and was deemed as INCOME. Projecting the FIT payment of £711 over 20 years (ignoring Index Linking for simplicity) then a minimum total of £14,220 could be anticipated. The annual consumption of mains electricity for 2014 was 4200 kWh – a saving of 1800 kWh (30 per cent) on an assessed (historical) annual usage of 6000 kWh. This amounted to an electricity bill saving of £221 for the usage of 1800 kWh at 12.28 pence per unit. This was defined as INCOME. (See Appendix Three for further detail). Thus the annual cost of mains electricity supplied was £515.76 with the cost per unit (kWh) at the time of 12.28 pence. The company Standing Charge has been ignored for simplicity sake. The total income for the property at Ceredigion was £932. Projecting this income over 20 years amounted to £18,640, and projecting the saved mains electricity figure of £221 over 20 years would result in £4,420. The above total income over 20 years has to take into account the initial capital cost of £7,100 for the solar PV system, which when deducted then gives a total income of £11,540. This again is a good investment on a capital expenditure of £7,100 especially when the saving in electricity consumption is also considered. With an income of £932 per annum the initial capital cost of the solar PV system is recoverable in 7.6 years.

Comparing the two systems

Over 20 years the projected income for Pembrokeshire amounted to £14,738, whereas Ceredigion (estimated) amounts to £11,540 – a difference of £3,198. The amount of mains electricity saved over 20 years being of the same magnitude, namely, £4,112 for Pembrokeshire, and £4,420 (estimated) for Ceredigion; a difference of £308 over 20 years. The system in Ceredigion having approximately twice the generation capacity of the system in Pembrokeshire, should produce approximately double the power of Pembrokeshire at 2056.2 kWh, and this proved to be the case with Ceredigion generating 4131 kWh per annum during 2014. The difference in income earned is also due to the differing FIT payments. The generation payment for Pembrokeshire was 43.3 pence for generation, with 3.1 pence for export, compared to 14.9 pence for generation and 4.6 pence for export at Ceredigion. It is interesting that the projected payback time for the solar PV system for Pembrokeshire is 6.9 years whilst Ceredigion at 7.6 years. This, as mentioned above, was due to the difference in the FIT generation. (See Appendix Three for additional detail).

Smart Export Guarantee (SEG)

The Smart Export Guarantee (SEG) pays customers for renewable electricity they have generated and exported to the National Grid. It replaces the Feed-in Tariff (FIT) scheme. The big energy companies have had to participate in the SEG since 1st January, 2020. As we have seen those who signed up to the Feed-in Tariff when it first launched in 2011 were paid a much more generous rate than those who signed up shortly before the scheme closed. Unfortunately and sadly SEG is not as generous as FIT and this is because SEG only pays for EXCESS electricity exported to the Grid, where the FIT paid for all the electricity generated by home solar arrays. Additionally, companies set their own SEG tariff prices so there will be a need to shop around for a competitive price, and at the time of writing prices ranged from 1p per kWh to nearly 6p per kWh. Nevertheless money would still be saved on the electricity bill as using the power generated from the solar panels means less being bought from the local electricity network. If a costly 'home' battery was installed more power could be saved from the electricity generated thus saving again on the home electricity bill. The scrapping of the FIT scheme from March, 2019 is myopic and needs reversing at the earliest possible moment. There needs to be greater encouragement, via household subsidies for the installation of solar panels. With the FIT scheme Government set the rates, whereas with the new SEG scheme companies set the rates making it much more cumbersome and complicated for the customer. For example, with SEG, you could be paid by suppliers in two different ways: a fixed rate or a flexible rate. A fixed rate SEG is a set amount of money for each kWh of renewable electricity exported to the grid, irrespective of the time of day, or a flexible rate which pays different amounts of money for renewable electricity depending on the time of day. My understanding of the scheme is the Department for Business, Energy & Industrial Strategy (BEIS) would like all fixed rate tariffs to be replaced by flexible rate tariffs, which rise and fall every half hour (based on wholesale electricity market prices). According to the Office of National Statistics (ONS) there were about 25 million households in the UK during 2017. Thus if 10 million of households each generated 4000 kWh per year with the installation of solar panels, this would amount to 40 TWh per year of solar generated power. This is nothing to be sniffed at, recognising that figures from the Department for Business, Energy and Industrial Strategy (BEIS) showed the UK generated 325 TWh for 2019. As an academic exercise it is interesting to note that if 20 million dwellings were fitted with 4 kW solar systems, then 80 TWh would be generated per year which would result in 25 per cent of UK demand.

Solar iBoost

Greater savings from home solar panels can be achieved as a result of the installation of a device called a Solar iBoost. The clever device has the ability to identify potential 'exported solar energy' and intelligently control and adjust the flow of energy, as an example, to the immersion heater of a hot water cylinder instead of the local power network. The device is manufactured in the UK by Marlec Engineering Co Ltd, Rutland House, Trevithick Road, Corby, Northants, NN17 5XY. (www.marlec.co.uk). During the period 2014 to 2018 a total of 6207 kWh of electricity was diverted to the hot water cylinder at the house in Ceredigion. As a consequence of having a Solar iBoost device fitted this resulted in a saving of £931 (at 15 pence per mains unit); of course, as electricity prices continue to increase so will the savings. During 2014 I was lucky enough to obtain a Solar iBoost unit for less than £250, thus the device has more than paid for itself and indeed continues to save money - what is there not to like. Therefore it is worthwhile and rewarding to consider this innovative energy saving device in greater detail, as every household that has solar PV panels fitted should invest in this very modestly priced option - I have no hesitation in recommending the device – it is a 'no brainer' as they say. Most solar panel systems are installed with a single generation meter which does do not differentiate between generation used at source and that exported to the local power network; the installed meter can only show the total solar generation. As such it should be clearly understood the householder will not easily be aware of energy exported at any one moment. From the householder's point of view, the energy will be lost to the power network – although this energy being available to another power company paying customer connected to the local power network. This is where the clever little unit called a Solar iBoost comes into its own, as the device has the ability to identify potential 'exported solar energy' and thus intelligently controls and divert the excess flow of energy, for example, to the house immersion heater, instead of the local power network. In effect this means that practically all of the power generated by a solar panel system can be used at source, resulting in considerable saving over time, and much to the delight of the householder. It should be noted though that when the hot water cylinder is full of hot water and the house is not demanding any power, then the excess electricity is exported to the local network. When export levels drop below 200 watts the unit switches off and waits until the export of 200 watt or more is restored, the unit then returns (switches) to water heating. To augment my solar panel system I had a Solar iBoost unit installed at the beginning of October, 2014 and have been more than satisfied with its operation having diverted on average 1511 kWh to the hot water tank immersion heater per annum - saving in the use of oil for heating the house hot water. This means the cost of the unit could be recovered in one year given good solar generation for any one year. In fact during 2015 the Solar iBoost diverted 1803 kWh which amounts to £270 at 15 pence per unit, thus could easily pay for the unit in a single year, thereafter it is all profit.

The benefits of the Solar iBoost are summarised as follows:

Solar iBoost fits quickly and neatly into an airing cupboard, simply wired between the existing fused spur and the immersion heater.

- It wirelessly receives information continuously from a sender device which activates the Solar iBoost to start water heating.

- The sender is battery powered so it is rapidly installed with its clamp in the utility meter box and no need for expensive wiring.

- The Solar iBoost intelligently controls and adjust the flow of energy to the immersion heater as the home consumption varies ensuring that only excess power is used.

- There is no need to change the immersion heater as the Solar iBoost works with any normal household immersion rated up to 3 kW.

- Solar iBoost displays real time and historical energy savings figures and LED symbols indicate the operating status.

- Simple timer programming enables Solar iBoost to work in harmony with Economy utility tariffs and a boost override switch means you can top up from the Grid any time extra hot water is needed.

Electrical energy diverted from the Solar iBoost did not affect FIT payments. For householders with a deemed usage contract from their FIT provider using Solar iBoost meant a greater or fuller usage of the free energy and still receive the export payment. When an export meter is fitted the benefits of the Solar iBoost can still outweigh the rising costs of water heating.

Heat pump

To further enhance the energy savings from the solar panel system fitted to the house in Ceredigion it was decided to invest in an air-to-air heat pump. The heat pump would use mains electricity far more economically in heating the house when required and effectively run *free* of cost when using solar generated electricity. Additionally, when employed in air- conditioning mode on a hot summer day, the heat pump used the *free* electricity from the solar panels. For those not familiar with air-source- heat-pumps, there is additional information later in the book which explains the technology, highlighting the advantages and disadvantages of heat pumps in much more detail. To summarise the benefits of my roof- fitted solar panels, there has been a reduction in mains electricity and central heating oil usage, the system continues to earn money from the Feed-in Tariff. The solar generation has been complemented and enhanced, by employing a Solar iBoost device and an air-to-air heat pump which helps to save even more energy and thus money. I should *stress* though that the combination of all these devices will not satisfy the heat requirement of the house in which I live, which is only satisfied by employment of an oil-fired central heating boiler.

Solar farms (parks)

There is a time and place for everything. Unfortunately due to our weather and latitude, large industrial arrays of solar panels are a *mindless* concept in the UK. On 11th January, 2018 at 13:15 GMT the total UK demand for electricity was 44.03 GW which came from CCGT (gas) at 24.22 GW (55.01 %), Nuclear at 6.46 GW (14.67%), Coal at 3.44 GW (7.8%), Wind at 3.27 GW (3.27%), with the balance from biomass, hydro, OCGT, pumped storage, French, Dutch and Irish Interconnectors, and **estimated solar.** Noting that day-time solar was **estimated** and is obviously zero at night. During the summer on 1st June, 2021 at 09:45 GMT total UK demand was 32.956 GW with only 6.53 GW (20%) coming from solar and 2.431 GW from wind. These are *actual facts* and can be viewed by anyone, anytime at www.gridwatch.co.uk. Credibility flies out the window with the proposed £400 million Cleve Hill Solar Farm, at Graveney, near Whitstable in Kent (at the time of writing it will be UK's largest solar farm). This project will cover 890 acres and is said to be capable of powering 110,000 homes. But what they do not volunteer, and it is vitally important, is that 110,000 homes will only have electricity when there is sufficient solar radiation. To keep the lights on any deficit will have to come from other sources and obviously will have to be provided by the National Grid. During the winter months especially around the winter solstice generation will be pathetic; a roof-mounted solar array with a total capacity of 4 kW will generate less than 1 kWh on a cloudy and cold December day. Heaven help the Grid (and all of us) if the network had to rely solely on large scale solar generation. Indeed, the whole project can be deemed doubly stupid when the electricity generated will be sent to the same sub-station that receives power from the London Array. The London Array is currently the world's largest offshore wind farm, having 175 wind generators and said to generate up to 630 MW. But again what is NOT volunteered is the generation of 630 MW is only accurate when the wind is at its optimum for maximum power generation. What happens on that freezing cold evening in the winter, when it is dark and there is no wind? If householders are *solely* dependent on these two very large arrays then they will be in *dire straits* – for *secure* power generation it is total madness! Then the whole concept of generating large scale electricity from solar radiation in the UK or the wind has a touch of the 'Emperor's New Clothes' about it. Such is the madness that wind farm operators have been paid almost £360 million to switch off their generators due to high winds or when generating too much power when the Grid does not need the power. Remember conventional power stations are directly under the control of power engineers, but electricity from solar (similar to wind) is at the mercy of the elements. It is utterly shameful that millions have been spent and wasted on these limited and very costly forms of power generation, which at the end of the day, dear reader, we all pay the price. Over a period of 7 years the average solar generation from a 4 kW solar panel system, mounted on the roof of a house in Ceredigion, produced 4000 kWh per year. The data was collected daily and recorded on a spreadsheet; thus daily, weekly, monthly and yearly figures were recorded. Ideally, if the system generated at full capacity for one year it would produce 35 MWh which would be extremely acceptable. Unfortunately due to cloudy skies and lack of the Sun at night, the system produced only 4 MWh and thus proved **11 per cent** efficient, which is normally expressed in engineering terms as the load factor (or capacity factor in the U.S.). Load factor is simply the ratio of the actual output (generation) and the output (generation) available x 100 over a period of time (one year).The data clearly demonstrated that solar power is far less efficient than the **25 per cent** electricity generated from the wind.

This begs the question to why government is pursuing such a LARGE SCALE *inefficient* and *costly* means of electricity generation - are Ministers innumerate as well as technologically and engineering challenged?

SOLAR FARM

A Dorset solar farm is said to be meeting the electricity needs of 60,000 homes in Bournemouth, that is, on a sunny summer's day. The solar farm first opened in 2014 and is expected to run for a substantial 25 years.

Chapel Lane Solar Farm, at Parley on the outskirts of Christchurch, cost £50 million to develop and covers 310 acres (125 hectares) – almost the area needed for 5,000 tennis courts. It is claimed the solar farm produces 51.3 MW of low carbon 'green' electricity when operating at *full* capacity. Well! That would indeed be excellent news if solar farms ran *continuously* at full capacity. Those who believe such claims are totally deluded and being led up the garden path by both the developers and Government. Developers know (or should know) that solar farms do not generate at night when household lighting, at the very least, is required, and that solar generation in the UK has an efficiency of just 11 per cent. The promotion of LARGE SCALE solar generation in the UK as presented by developers is a classic case of *smoke and mirrors* – do not be fooled!

Environmentalists claim solar farms are a green and sustainable form of generating electricity - this claim is skin deep and delusional and we have shown that large scale solar parks in the UK are woefully inefficient in producing large amounts of electricity. There is also the consideration of the majority of panels being made in China with their transportation, which adds to global carbon emissions. The panels also have a life span of less than fifty years and there appears to be little consideration given to the eventual disposal of thousands and thousands of these panels. It has been shown that roof-fitted solar panels are relatively green, but large scale solar panels are certainly not. They are made using rare earth elements, the supply of which has both capacity and political issues. Solar panels are manufactured at 2000°C, a temperature so high it requires fossil-fuel power - solar cells are formed using silicon dioxide as a raw material, which requires purification and as such involves heating Silicon dioxide in an electric furnace. The solar panel consists of a solar cell, glass casing, metal frame, and wires which allows the flow of current from all the cells. But as we shall see, *'flexible thin film solar photovoltaic cells'* when fitted to buildings, are much more environmentally and aesthetically acceptable, and are the way forward.

Flexible Thin Film Solar Photovoltaic Cells

It is said that buildings are responsible for approximately 40 per cent of the UK's energy consumption and associated carbon emissions. Therefore it can be argued, there is a lot of sense, if properties could be turned into small, environmentally friendly, power stations without detracting from the aesthetic appeal of the property.

This is where the fixing of what is known as *flexible thin film solar photovoltaic cells*, directly onto roofing would make a lot of sense, with the cells being *suitable* for commercial, industrial and residential roofs. Indeed, other buildings, such as stations, stadiums and churches, that are due for re-roofing could also benefit from the integration of the 'aesthetically sympathetic' look of thin film solar cells. Additionally the cells have zero wind impact and are much lighter at less than 3 kg per square metre as compared to conventional PV solar panels at 1520 kg per square metre. These attribute obviously opens up the market to a wide range of use, as buildings with little load sparing capacity can be considered for safe application. Flexible solar panels employ CIGS technology (Copper Indium Gallium Selenide) thin-film solar cells to convert sunlight to energy - made from either lightweight crystalline cells or thin film coated in plastic that can be bent or curved to fit complex structures.

The conventional roof-fitted solar panels (crystalline silicon PV cells) are fragile and not flexible, hence require glass to protect them from damage from external forces and internal thermal and mechanical movements, and an aluminium frame is required to support the glass – and as mentioned above results in a heavy and bulky solar photovoltaic panel. The rigidly fixed traditional crystalline based photovoltaic systems use heavier and bulkier solar photovoltaic panels, and are usually mounted on top of roofs or other spaces after the construction of the building. Unfortunately this method of attachment cannot claim to be aesthetically attractive and as such have not always been popular. To do away with the unsightly crystalline silicon PV cells it is necessary to do away with the glass and aluminium frame. This means the photovoltaic cell itself will need to be robust enough to be attached directly to a roof and therefore will need to be flexible. Manufacturers say the advantages of CIGS technology compared to traditional crystalline technology are less bulky, more flexible and light weight, more durable with better shade tolerance, performs better in low light conditions, requires fewer materials, less energy to manufacture and requires very little maintenance. Unlike silicon, thin-film photovoltaic cells are relatively insensitive to shading. This means that shadow, dirt or surface damage does not affect the overall production, as is the case with silicon photovoltaic panels. This makes CIGS a very effective solution in integrated urban and industrial solar cells – having the ability to turn buildings into power stations without making them look like one.

Ministerial decisions

Technological and engineering challenged politicians such as Chris Huhne, Ed Davey and Amber Rudd have a lot to answer for. Future generations will find it hard to forgive the lethargy in allowing uninformed politicians to create such an appalling energy policy there is today. It justifies reiterating that Ministers such as Amber Rudd were educated at

Cheltenham Ladies' College, an independent school in Gloucestershire, and read history at Edinburgh University. Therefore with the greatest respect, I am not aware if Amber Rudd has any education in the necessary science, technology, electrical and power engineering – obviously education, knowledge and skills in these disciplines would be of great value in making much more informed decisions. What are we to make of the disgraced Chris Huhne, former Energy and Climate Change Secretary of State, who together with Vicky Price were sentenced for perverting the course of justice in Southwark Crown Court. This Liberal Democrat was the influential advocate of carbon emission reduction in the Conservative led Coalition of 2010-2015; Mr Huhne (at the time of writing) is the European Chairman of Zilkha Biomass Energy (HQ at Houston Texas), a company that makes wood pellets for converted coal-fired power stations. It was observed at the time that Tory MP Jacob Rees-Mogg said of Mr Huhne: "This man served a prison sentence so he is not necessarily a model of uprightness. Wood pellets are yet another clean energy scheme that has turned into a racket." Britain's largest power station, Drax in North Yorkshire, received more than £450 million in subsidies in 2015 for burning biomass, which were mostly U.S. wood pellets. But surprise, surprise, during December, 2013 Energy Secretary Edward Davey visited Drax Power Station to celebrate Drax becoming one of Europe's biggest pellet burning electricity generators! The Energy Advocate (online) research suggests that the use of wood pellets as a source of power is worse for the environment than the use of coal – and is being subsidised by British consumers. Acres of trees have been felled in the U.S. and shipped over to the Drax power station in North Yorkshire to be burned as biomass, which has been promoted as 'cleaner and greener' than coal. However, research by British academics says burning these wood pellets produces more carbon emissions than coal, with the transportation over thousands of miles in dirty diesel ships adding to emissions - the track record of former Energy Ministers is nothing to be proud of, quite the opposite. During February, 2010 the Energy and Climate Change Secretary, Ed Miliband, said that from 1 April, 2010 households with approved schemes will be paid for the electricity they generate, even if they use all of it themselves. When he announced the tariff' Ed Miliband said the guaranteed income would be a big incentive for householders to make the move to low carbon living, and the Feed-in Tariff will change the way householders and communities think about their future energy needs, making the payback for investment far shorter than in the past. It was also said, that importantly, future payments were guaranteed for 25 years and would be linked to inflation. According to government figures at the time a typical 2500 kW solar installation could offer a homeowner a reward of up to
£900 and save them £140 a year on their electricity bill; despite the payback, the upfront cost put off many householders, with the average

price of the installation of solar panels then around £10,000 to £12,000. It was also interesting that the energy regulator Ofgem, warned the nation to expect 20 per cent electricity price hikes by 2020, and warned future supplies were in jeopardy, investors in solar panels will have the added benefit of being a net provider of electricity, and largely insulated from future price hikes that could see household bills top £2,000 a year by 2020. It is vital for the country to have a Secretary for Business, Energy and Industrial Strategy, who has the sense and ability to understand the necessary science and engineering, thus enabling a halt to the spread of both large wind farms and large solar parks. Indeed, Ministers should be offering far better solutions and inducements. Roof-fitted *flexible thin film solar photovoltaic cells*, coupled with the reliable and sensible use of tidal energy should be vigorously pursued. Surely Ministers are aware that the second highest tidal range on the planet takes place in the Bristol Channel
– a perfect location for tidal lagoons and other means of tidal generation. See tidal energy later in the book.

The prime motivation for my fitting solar panels was deemed as a way in which to reduce the mains power bill, which of course, was much encouraged by the then, generous FIT payments, and has proven to be a sound investment. I have to be honest and admit the fitting of roof-solar panels was not driven by an environmentally-friendly aspiration; having said that, my wife and I try and do our best for the environment, having both been born during the Second World War. The deprivations of the war years and indeed afterwards, with shortages, rationing and power cuts coupled with the hard winter of 1947 taught valuable lessons. Indeed, the *University of Life* has conditioned and taught us both not to spend beyond our means, to save for that rainy day, to grow our own vegetables, to recycle and repair as much as possible – we abhor waste! With hindsight (a marvellous thing) I regret that I did not invest in a 4 kW system in the early days of roof-fitted arrays when the FIT was generously paying more than 50 pence per unit - not to mention the saving on household mains electricity bills. But back then installation costs were high, and the *continuance* of such generous payments from Government seemed too good to be true, even though Government assured they would honour such payments at the time of installation. Thus caution kicked in and I settled for a 2300 kW system. Unfortunately, since the FIT scheme that was introduced during a Labour Government on 1st April, 2010, the initial generous payments have steadily diminished, until they were scrapped by a Conservative Government at the end of March, 2019, with the scheme closing to new applicants. A far less generous scheme, the Smart Export Guarantee (SEG), has recently been introduced and which poses a conundrum as our leaders wish us to use more GREEN energy – surely an attractive subsidy would do this - or perhaps they think we are all green?

CHAPTER SIX

'It's easier to fool people than to convince them that they've been fooled'

Samuel Langhorne Clemens (1835-1910)
(Mark Twain)

ELECTRIC VEHICLES

Having been schooled and trained in electrical engineering (specifically telecommunications), I am naturally attracted to the concept of Electric Vehicles (EVs), especially when appreciating that obnoxious exhaust gases will not be emitted to the atmosphere by such modes of transport. With the absence of petrol and diesel fuelled vehicles, polluted towns and cities will certainly have their environment improved. Although it is recognised that electric vehicles are not totally green. For example, life- threatening substances such as particulate matter will still be a threat as vehicle tyres and road surfaces will still wear down - but even more so with EVs, as battery-powered cars, by their very nature, are much heavier than petrol or diesel versions due to the housing of a heavy battery – the weight of which, resulting in more wear and tear. Particulate matter is everything in the air that cannot be defined as a gas, and this includes natural sources like pollen, sea spray and desert dust. It also includes human made sources like smoke and dust from exhausts, microplastics from brake pads and tyres; particulate matter can travel large distances. Consider the millions of vehicles in the UK with their health threatening exhaust emissions. The tiny particulates of carbon found in air are the direct consequence of this traffic, and are known as PM2.5 - these are particles less than 2.5 microns wide - a micron is a millionth of a metre. The particulates can bypass the mucus in our airways that trap dust and pollutants so rendering this 'filtering system' ineffective and allowing pollutants to enter the body. This is particularly dangerous for those suffering from heart and lung problems. Scientists argue and warn that the air pollution levels are so high on Oxford Street, London, that just by spending two hours there can cause significant stiffening of the arteries.

Indeed a group of MPs called for planning guidelines to be changed so schools and care homes could no longer be built near pollution hotspots; according to a study at St George's hospital, South London, one in fifty heart attacks that lead to admissions at London hospitals may be triggered by air pollution. It was very sad and shameful day when a Southwark, London coroner ruled in December 2020 that air pollution and illegal levels of nitrogen dioxide contributed to the death of a nine year girl who lived 80ft from the South Circular road in Lewisham, south-east, London. What are we to make of the weasel words from Ministers who pledged to set aside an extra £6 million for local authorities to improve air quality and raise public awareness, and new legal limits, which could be brought in by October 2022 following a consultation? Indeed, Environment Secretary George Eustace said that the new targets would be *informed* by more stringent World Health Organisation standards. You really have to wonder in what parallel universe Ministers exist with their insulting weasel words and a promise of £6 million for local councils. Just how far will this limited amount of money go? Exactly what *projects* do they envisage for cleaning up the air? How patronising the claim to enhance public awareness, when, dear reader, we are all well aware of the curse of 21st Century pollution.

It is vital for action NOW and not vacuous promises and the prospect of jam tomorrow!

It would appear the lessons of the last two centuries are soon and conveniently forgotten by politicians - when Londoners, back then, suffered from extreme Smog - this is a type of air pollution consisting of smoke and fog (commonly known then as pea soup fog). It was a familiar and serious problem for Londoner's from the 19th century to the mid-20th century, due to the burning of large amounts of coal in homes and factories. Smog consists of soot particulates from smoke, sulphur dioxide and other components. Smog itself is simply airborne pollution which may obscure vision and cause various health conditions. It is caused by small particles of material which become concentrated in the air for a variety of reasons. Commonly, smog is caused by an inversion, in which cool air presses down on a column of warm air, forcing the air to remain stationary. The threat to health also rears its ugly head when a high pressure system dominates over the UK and the air flow is from Europe, coupled with dust blowing up from storms in the Sahara. This combination resulted in extreme air pollution (smog) over London and other parts of the country on Friday 10th April, 2015 with the temperature reaching 21.9^0 C at St James's Park in London and 21.3^0 C at Heathrow Airport. People with lung and heart conditions were warned against exercising outside. Make no mistake as unless this situation is addressed purposefully, it will result in an ever increasing debilitating scenario with normally healthy people beginning to succumb to breathing difficulties.

TRAFFIC AIR POLLUTION

Modern smog (Photochemical smog) is a type of air pollution, derived as a result of vehicular emission from internal combustion engines and industrial fumes that react in the atmosphere with sunlight to form secondary pollutants that also combine with the primary emissions to form photochemical smog. The atmospheric pollution levels of cities such as Mexico City are increased by inversion that traps pollution close to the ground - it is usually highly toxic to humans and can cause severe sickness, shortened life or prove fatal. Photochemical smog is a unique type of air pollution caused by reactions between sunlight and pollutants like hydrocarbons and nitrogen dioxide. Although photochemical smog is often invisible, it can be extremely harmful, leading to irritations of the respiratory tract and eyes. Nitrogen oxides are a group of gases that are mainly created from burning fossil fuels. When the gas reacts with others in the air, it can create nitrogen dioxide (NO_2). Nitrogen oxide emissions in the UK come from: 35 per cent road transport, 22 per cent energy generation, 19 per cent industrial combustion, and 17 per cent from other transport, such as rail and shipping. In regions of the world with high concentrations of photochemical smog, elevated rates of death and respiratory illnesses have been observed. Some of the particulate matter in the air can oxidize very readily when exposed to the Ultraviolet (UV) spectrum. It does not have to be that sunny for photochemical smog to form as UV light can penetrate clouds. The pollutants released through human activity in this situation are known as 'primary pollutants' and they include sulphur dioxide, carbon monoxide, and other volatile organic compounds. When these compounds interact with the Sun, they form 'secondary pollutants' like ozone and additional hydrocarbons. It is very sobering and frightening to consider all the vehicles we encounter daily on the roads of the UK that are burning oxygen and spurting out obnoxious pollutants, and then contemplate all the vehicles on the roads globally in countries such as China, France, Germany, Italy, Spain, Africa, India, Australia and the North and South American continents – not a day or night passes without tons of pollutants pouring into the atmosphere. This increasing pollution begs the question as to whether electric vehicles are the answer to this nightmare – can the country provide, for example, the infrastructure plus the adequate and reliable energy required, not to mention the high cost of purchasing an electric vehicle! The workings of the electric vehicle are quite elementary as the driving wheels are connected to one or more electric motors, which are then simply powered by the vehicle battery. When the accelerator pedal is pressed the battery instantly feeds power to the electric motor, turning the wheels and moving the vehicle. The distance the vehicle can travel is directly proportional to the amount of energy the battery can store, before requiring charging. It should be noted that electric motors can also work as electrical generators, such that when the driver's foot is taken off the accelerator pedal, the vehicle will slow down by converting its forward motion back into electricity slightly improving the vehicles range.

As mentioned earlier that as a result of tyre and road surface wear it is recognised that EVs are not completely environmentally friendly, but then appreciation of the manufacture and the use of lithium-ion batteries also need to be considered. As the name suggests lithium-ion (Li-ion) batteries do not contain lithium metal, but they contain ions - an ion is an atom or molecule with an electric charge caused by the loss or gain of one or more electrons. The batteries for EVs are reliant on cobalt mining. Most of the cobalt in the world is mined in Africa, specifically the Democratic Republic of Congo (DRC). It is mainly produced as a by-product of platinum and nickel mining operations. Cobalt is a key component of lithium-ion batteries that power mobile phones, laptop computers, tablets, and it is also used in several military and industrial applications. About 60 per cent of the world's cobalt comes from DRC mines, where the metal is sometimes dug out by hand in unregulated conditions, often by child labourers who risk their lives for about £1.50 a day. So it can be argued that our so-called clean technology comes at a price in human suffering - a single electric car needs between 6-12 kg of cobalt, with of almost 120,000 tonnes a year anticipated by 2030. In June, 2019, a team of scientists wrote to the Committee on Climate Change, warning of the huge increase in the production of cobalt and other raw materials that will be necessary if all of the UK's estimated 40 million cars were to be replaced with electric vehicles by 2050.

In scrapping fossil-fuelled cars it should be recognised that each vehicle apart from its body and engine etc, will have at least four tyres to dispose of bearing in mind that historically, waste tyres have been a nightmare for society and largely ending up in landfill sites where they occupy a lot of space; although legislation, has led to huge increases in recycling rates for tyres, thus becoming more eco-friendly. Metal rims are sometimes still attached when they are received by a recycling company. These are easy to separate and can be refurbished and re-sold or treated as ferrous scrap metal – the rest of the remaining tyre will normally be made of a mixture of steel, textile and rubber, although there are often traces of oil and other chemicals present as well. Tyres can sometimes be re-treaded and re-used but sometimes, the only option is to recycle them, particularly if they are badly damaged. Fortunately tyre shredding is largely increasing. The metal is recycled normally, whilst the rubber crumb is sometimes used to make sports surfaces and safety mats for play areas. It also has many other uses, ranging from carpet underlay to the production of rubberised asphalt for roads or even as a fuel. To assist on used tyre recovery issues the Used Tyre Working Group (UTWG) was formed in 1995 to act as a link between industry and Government. Thus with the advent of electric vehicles over a relatively short span of time, over 160,000,000 tyres will have to be dealt with, being re-used or scrapped. Therefore I wonder if the UTWG and Government are liaising and preparing for this upsurge of used tyres. EVs will not have an exhaust pipe emitting obnoxious substances to the atmosphere, but there will be particles created from tyre and break lining wear, which breaks down into fine dust and it certainly does not disappear into thin air! It should be noted that apart from the regenerative braking system, which relies on the car's electrical system to operate, hybrid and EV brakes are entirely conventional – similar to the disc and drum brakes fitted to any other car. Heavy road freight vehicles have a huge impact on our roads, which were never designed, nor intended for such traffic - damaging pavement structures, city streets, local roads, and county highways. Lorries travelled 17 billion miles during 2017, and it is staggering that about 100,000 foreign heavy goods vehicles (HGVs) make 1.5million trips to Britain every year. It is disturbing to think of all that wear and tear on tyres, road surfaces and the tiny particulates as a result – to all this we will now have to accommodate the heavier electric vehicle. A study by Utah State University, U.S. (April, 2021), looked at sources and locations where microplastics were more concentrated, and found that roads produced 84 per cent microplastic waste found in the atmosphere. The study discovered microplastics from the land were present in ocean water, and came to a conclusion that microplastics are spreading throughout the atmosphere. The study discovered the microplastics in the atmosphere originated from road dust, which were specifically left by tires. Areas, where the microplastics were most accumulated, included Europe, Eastern Asia, the Middle East, India and the U.S. Based on the size of the microplastics, the particles remained in the atmosphere from barely an hour to 6.5 days at times, which allowed the particles to move to a different continent.

In 1953 the railways were carrying slightly more tonne-mileage than the roads, but by 1979 there were five times as much tonne-mileage on the roads than on the railways. It was outrageous, shameful and extremely short sighted when Dr Richard Beeching sparked uproar in the 1960s and closed 4,500 miles of railway line and 2,128 stations to save money. The Beeching cuts (also known as the Beeching Axe) were a great reduction of route network and restructuring of the railways in Britain. Dr Richard Beeching was responsible for two reports, namely, The Reshaping of British Railways (1963) and The Development of the Major Railway Trunk Routes (1965), and written by, and published by the British Railways Board as it was known then. The Beeching cuts were myopic and extremely savage when dozens of branch lines that linked villages with market towns were rated egregious - indeed, loss-makers to be culled, along with great chunks of mainline. The railways that had helped Britain become an industrial power, but which were now losing money, were cut back brutally. The car would replace the train, Beeching decreed. In doing so, he ushered in an era of vast motorway expansion and cheap motorised transport resulting in the vast pollution and traffic problems we have today. If Beeching and other transport planners of the day had had their way, only a rump of inter-city lines would have been left. But today the makeup of Britain's transport looks very different from the one envisaged by Dr Beeching. Rail passenger figures have risen and commuter trains are crammed; young people are deserting the car for the train, and Britain's railway bosses are struggling to meet soaring demands for seats. The legacy of Beeching with dug-up lines, sold-off track beds and demolished bridges, has only hindered plans to revitalise the network, revealing the dangers of having a single, inflexible vision when planning infrastructure. Of course, a number of lines were loss-making but nowhere near the number Beeching axed to our cost. All large freight should be catered for on the railways as a large freight train can replace many dirty diesel HGVs. Electrification of the rail network would greatly help reduce atmospheric pollution from vehicles and hazardous substances are *safer* being transported by rail, The infrastructure should encompass rail to numerous centres, freeing up the motorways, with onward distribution from these centres to supermarkets and shops by smaller lorries and vans.

Perhaps my *attraction* for an integrated road and the railway network, was *encouraged* at the age of 14 years, when straight from school, I was taken on as an office boy to the Personal Secretary of the Divisional Manager of what was then British Road Services (BRS), at their offices in the Andrews Arcade, Queen Street, Cardiff. The Company (BRS) was established in 1948 and was part of the British road transport company formed by the nationalisation of Britain's road haulage industry, under the British Transport Commission, as a result of the Transport Act 1947. By the 1960s the company was made up of four main operating areas: British Road Services Ltd., BRS Parcels Ltd., Pickfords & Containerway, and Roadferry. It is interesting to note that during the Beeching cuts in the 1960s the Conservatives led by Harold Macmillan were in power from 1957 to 1963. This was followed by a Labour Government under Harold Wilson from 1964 to 1970. Both Governments were complicit and responsible for the decimation of the railways. Although the cuts started off under a Tory Government, it was continued under a Labour Government and therefore Labour cannot wash their hands of any responsibility. Harold Wilson's Labour government continued with the policy of closing uneconomic lines, but it came under increasing pressure from backbenchers. Therefore in 1966, a White Paper on Transport Policy identified economic utility as the major objective of railway policy. This resulted in a revised railway network plan with 3,000 miles of additional track surviving the Beecham cuts. Following a policy review in 1967, the Transport Act of 1968 made provision for major capital reconstruction on the railways and deficit relief. Certain routes fulfilling an important function that were not financially viable were subsidised. In 1969 British Road Services was renamed the National Freight Corporation. In 1980 the assets of the National Freight Corporation were transferred to the National Freight Company. Under the Conservative government of Margaret Thatcher in 1982 it was sold to its employees in one of the first **privatisations** of state-owned industry under the name the National Freight Consortium.

With regard to the environment and general health there is much to commend the change from diesel and petrol cars to that of the electric version. But on the other hand there is much to consider in bringing about this *monumental* change. When politicians and others lecture on the virtues of electric vehicles the song 'The Impossible Dream' quickly springs to mind, as these myopic people have certainly given little thought to where all the extra electrical energy will come from, let alone how the current network will cope without cables overheating, transformers and switch gear burning out. Do the ill-informed and myopic politicians not appreciate the damaging burden that will be put upon the existing power network – a network that was not designed for such electrical loadings – major and costly upgrades will be needed. Additionally, if these dullards are hoping that wind and solar will provide the extra energy, even if the network could cope, it will be tears before bedtime.

Power for EVs

A Nissan Leaf car (cost £26,845 new at November, 2020) has a 40 kWh capacity battery and can travel 168 miles on a full charge, whereas the Nissan Leaf E+ has a 62 kWh battery capacity and can travel 239 miles on a full charge. Nissan offer two options for charging, namely the 7 kW charger at home, office or on the road, and secondly, the 50 kW Chademo rapid charger, which will charge the Nissan Leaf in 60 minutes or the Leaf E+ in 90 minutes from 20 per cent to 80 per cent. To fully charge one Nissan Leaf car with a fully discharged battery (capacity of 40 kWh) at 7 kW will need 5.7 hours of charging. (It should be noted though that in practice, after the initial full charge, the battery should not be drained to a fully discharged state). Batteries are seldom fully discharged, and manufacturers often use the 80 per cent **Depth-of-Discharge (DoD)** formula to rate a battery. This means that only 80 per cent of the available energy is delivered and 20 per cent remains in reserve. Nissan recommend charging the battery at 80 per cent to preserve battery life, and a battery is restricted to 80 per cent charge only if charged through a rapid charging station (50 kW) to preserve battery life. Numerous factors determine how far an EV will travel on a full charge. Driving on a wet winters evening with lights, wipers and heater on will certainly lessen the range of the car.

In *reality* 40 million EVs will not be charging *all* at the same time, or charging a *flat battery to a full battery*, but mainly charging between 20 per cent and 80 per cent of battery capacity. Additionally, all 40 million EVs will hardly be charging their batteries during normal Grid peak demand. It could be argued that EVs will create their very own peak when electricity is cheap, that is, *overnight* when most people are at home from work. There would possibly be a second peak on weekdays in the morning, between 0900 hrs and 1200 hrs, due to charging at work and at slow/fast public charge points. It is envisaged that that will be three main types of EV charge points – Slow, fast and rapid:

Residential: Charge points located at or near EV drivers' homes having a rated capacity of 3-7 kW.

Work: Charge points installed in workplaces, having a rated capacity of 3-22 kW.

Rapid Public: An accessible charge points to the public with a charging capacity ≥43 kW.

Complementing the above will be the Slow/Fast Public, publicly accessible charge, excluding those classified as Work or Residential, with a charging capacity ≤22 kW.

During January, 2022, there appeared an advert for EVs in a local paper which included charging and driving tips for the most efficient use of an electric vehicle. The advert advised to charge the battery to a maximum of 80 per cent as it would extend the lifetime of the battery – and that the healthiest state of charge of the battery is between 30 per cent and 80 per cent. The advert further stated that rapid charging sessions would cause high currents and high temperatures which will strain the battery, and said there would be no big problem with supercharging multiple times per week, but try to use them only when needed. Charging the last 20 per cent of the battery could take almost as long as charging to the first 80 per cent. Attempt to decelerate as much as possible by using regenerative braking as it will minimise energy consumption. Use of the EVs eco-mode will make acceleration gentler and save battery life. To remove roof racks or rear racks when not in use, as they will shorten the vehicles driving range. Improperly inflated tyres will increase energy consumption thus draining the battery. Fast acceleration might be nice, but unfortunately it eats away battery energy. Try choosing energy-efficient routes in the EVs navigation-system. Thus electric vehicles will not come without an irritating level of hassle.

Ofgem has called for incentives to encourage people to charge their EVs outside of peak hours, which they say would increase the number of cars currently supportable by the country's electricity network by 60 per cent. This is very similar to the National Grid's own opinion, which is that flexible charging would halve the estimated additional generation needed to manage the demand. National Grid on their website (www.nationalgrid.com) say there could be 36 million electric vehicles on the roads by 2040 and through *smart charging technologies*, charging vehicles at off-peak times and through vehicle-to-grid technology, the increase in peak demand from vehicles could be as little as 8 GW. The author certainly challenges the figure of 8 GW (unless consumers are being purposefully cut-off), demonstrating the figure to be extremely low due to the following reasoning: At the time of writing there are 39,995,000 vehicles registered for use on UK roads including cars, vans, taxis, buses and trucks. That is a considerable number of petrol and diesel vehicles to be disposed of and replaced by electric vehicles. The Government backed organisation 'Smart Energy GB' which is tasked with informing Great Britain about the benefits of the Smart Meter rollout has bizarrely stated in one of its adverts, 'It is your Smart Meter that will communicate with other tech such as electric cars or smart washing machines so they know when energy is cheapest and should charge or turn on. Electric cars will eventually overtake petrol and diesel and their batteries will allow homeowners to store electricity and sell it back to the Grid, but only if we have upgraded to a smarter system'. Really! More misleading claims from Smart Energy GB - how will this work in practice? If electric car owners are to sell electricity back to the Grid then obviously they will have to be connected to the electricity network, draining the car battery of energy – talk about having your cake and eating it! If Smart Meters are to communicate with cars, does this mean this can only happen when the car is plugged into the home supply, or is it the intention to have a radio link between the Smart Meter and the electric car so the driver, when on the road, can speed to the nearest charging point? Then we have the multi- million dollar question of how much extra power will the Grid have to supply for nearly 40 million EVs and can the existing network *really* handle the extra load, not to mention all the additional cost and cabling to various premises and supermarket car parks to provide charging facilities. To appreciate the magnitude of the problem it will be instructive to consider the annual and peak demand of electricity in the UK - as an example, the total UK annual electricity demand was 346 TWh in 2019, and the peak demand for electricity was 47.275 GW.

Obviously 40 million new vehicles will not suddenly appear on the highways, but will increase steadily over the years. At the time of writing the Prime Minister, Boris Johnson, hailed a 10-point plan for a 'Green Industrial Revolution' with the electric car at its centre – the sale of all new petrol and diesel vehicles to be banned from 2030. I wonder if this means that my stand-by petrol driven home generator will then become a museum piece! Living in a rural environment it has proved its worth – saved a Christmas dinner a number of years ago when the local mains supply failed!

National Grid claim, '*The growing use of electric vehicles could increase electricity peak demand by 3.5 GW in Britain by 2030, but peak electricity demand could even rise by as much as 8 GW by 2030 without "smart charging" during off-peak hours and 18 GW by 2050*'.

If the National Grid figure of 3.5 GW is to be believed then that means an additional large power station that is capable of supplying 3.5 GW (3500

MW) will be required - and two such stations if the 8 GW figure is realistic. My own calculations clearly indicate the National Grid figures are tongue-in-cheek, implying the Government and its advisors of not carrying out any meaningful research. If the country has to accommodate 40 million EVs, my calculations, based on the Nissan Leaf car specification (see Appendix Four), show that if 10 percent of total EVs are charging simultaneously, then an additional 28 GW will have to be found. This will require twenty two large CCGT power stations each of 2 GW (2000 MW) capacity with a load factor of 60 per cent. Should my predicted figure for vehicles charging simultaneously be considered high, and that 5 per cent (2 million EVs) is a more reasonable figure, then eleven such power stations will still be required. But recognise dear reader that if a quarter of all EVs should charge simultaneously, then the demand would be 70 GW and that would be double that of the 47.275 GW peak demand during 2019.

Therefore the obvious conclusion is that there will be a considerable increase in UK power demand if 40 million EVs ever become a reality.

For the year 2019 the UK annual power consumption resulted in a 346 TWh. If 40 million EVs become a *reality*, then my calculations indicate there will be a need for an additional 95.2 TWh, resulting in a new annual consumption of 441.2 TWh. This additional annual requirement would call for nine large CCGT power stations each of 2 GW (2000 MW) capacity with a load factor of 60 per cent. The Department for Business, Energy and Industrial Strategy on 5 July, 2019 (Digest of UK Energy Statistics 2019) stated the domestic sector remained the largest electricity consumer in 2018 at 105.1 TWh, while the industrial sector consumed 93.0 TWh, and the service sector consumed 96.6 TWh. The average household according to Ofgem uses 3100 kWh per year. Researching the Internet and according to a number of sites it appears the change to EVs will result in each household using an estimated extra 2000 kWh to 3000 kWh per year in electricity.

Please note that during the 2021 coronavirus pandemic the UK electricity demand was 330 TWh and the peak demand was 47.107 GW.

RELIANCE ON ELECTRICITY FROM WIND OR SOLAR IS BIZARRE

According to the Office for National Statistics (ONS) there were approximately 25 million households in the UK during 2017. Therefore to satisfy EV charging each household will require a minimum additional 2000 kWh per year, and the UK will need to generate an extra 50 TWh per year to charge household EVs. This will mean building an additional 5 CCGT large power stations, each having a capacity of 2 GW (2000 MW) and load factor of 60 per cent. It should be recognised that a 2 GW CCGT power station costs in the region of £1 Billion to build, thus 5 such new stations would cost £5 billion. If the true demand figure is closer to 3000 kWh then nearly 7 new stations at a cost of £7 billion will be required. If this extra power is to be satisfied by wind farms with each individual wind generator having a maximum capacity of 2 MW, and an efficiency of 26 per cent then an extra 11,000 wind generators will be required to provide the additional 50 TWh per year. But, of course, if the additional power required is found to be 3000 kWh then an extra 17,000 wind generators will be needed. It is extremely important to remember that wind generation, due to its very nature, is variable and as such can be generating maximum power when minimal requirement is needed by EVs, but more importantly can be generating minimal power when EVs need it most. As a point of interest it is enlightening to note that an Internet search revealed that by comparison the U.S. household uses 12,300 kWh per year, Canada, 11,000 kWh per year and Australia, 7,000 kWh per year, while the UK used 4,200 kWh per year.

Electric Vehicles employ a lithium-ion battery which is rechargeable. The Electric Vehicle Battery (EVB) also known as a traction battery is used to power the electric motors. The batteries have a higher charge capacity (energy density) than typical lead-acid or nickel-cadmium types, which means that battery manufacturers can save space, reducing the overall size of the battery pack. It should be noted that Lithium-ion batteries in mobile phones begin to wear out after only a couple of years, and during that time the phone might be fully charged and discharged hundreds of times. But every charge cycles counts against the life of the battery, and after approximately 500 full cycles, a lithium-ion phone battery will begin to lose a significant part of the capacity it had when new. While that might be acceptable for a phone, it's not good enough for a car designed to last many thousands of miles, thus EV manufacturers have ensured their electric car batteries last longer. To the extent that these car batteries are 'buffered', meaning that drivers cannot use the full amount of power they store, reducing the number of cycles the battery goes through. Together with other techniques such as clever cooling systems, means that electric car batteries should give many years of trouble-free life. The battery on an electric car is a proven technology offering many years usage before replacement is necessary. Many manufacturers provide a warranty of up to 8 years or 100,000 miles before replacement is necessary; a new 40 kWh battery in 2016 would have cost just under £10,000, but it is claimed that prices are set to fall by 2030. However, the current prediction is that an electric car battery will last from 10 to 20 years before they need to be replaced. Unless there is an improvement in battery charging times, 30 or 40 million battery driven EVs will not materialise - especially when the necessary network/infrastructure and charging facilities are not there to accommodate them! An aerial view of our towns and cities will quickly identify the numerous terraced houses and tall tower blocks of flats – are the streets and roads to be entangled with a jungle of cables charging all the proposed nice, new EVs - consider the scenario with charging points for 40 million cars - streets and roads becoming a labyrinth of cables – a tangled and impenetrable suburban spider's web - cables dangling down from high rise flats, and criss-crossing in streets that have terraced houses. In the *real* world how many EVs can be accommodated with street/road charging facilities? National Grid on their website (www.nationalgrid.com) state that, 'Over 40 per cent of people in England and Wales do not have off-street parking'. If it is envisaged that charging cables will indeed run across Public Rights of Way such as pavements, they will prove to be extremely hazardous and dangerous, especially for the disabled and elderly. Will the lawyers have a field day making a fortune from illegal trip hazards? What safeguards, if any, could be put in place to avoid such bone shattering accidents? The existing local networks were not designed and built for the extra loading that EVs will inevitably bring, even 20 million EVs will place serious demand on the network and unless the necessary upgrades are implemented the consequences will be overheating cables, switchgear and transformers, resulting in plant failure followed by brownouts and blackouts. Could this network *'Sword of Damocles'* possibly be one of the reasons why Government and the power companies are pursuing the fitting of Smart Meters, as they will offer the potential of disconnecting consumers from a remote computer when parts of the network are threatening over-demand for electricity – EV owners will not be happy bunnies if there is no power available to charge their vehicles when required. Then, of course, there is the high price of EVs and unless there is a significant decrease in cost, how many people will be able to afford to purchase a new EV?

Unless all the aforementioned obstacles can be overcome it is all nonsense and is exacerbated by the fact that road transport is not the worst culprit regarding atmospheric emissions.

Shipping

We forever read in the media of the emissions from land transport, but very little about emissions from transportation on the oceans and seas. How many readers fully appreciate that at January 1st 2021 there were around 56,000 merchant ships trading internationally. The vast majority of these ships are no longer sailing or coal-fired. Since about 1960 most new ships have been built with diesel engines. The last major passenger ship built with steam turbines was the Fairsky launched in 1984. This was a single-class Italian-styled liner operated by the Sitmar Line, and was primarily used from 1958 to 1972 on the migrant route to Australia from the UK. Most modern ships employ a reciprocating diesel engine as their prime mover, due to their operating simplicity, and robustness. I deem it excellent and congratulate the Daily Mail for publishing during 2016, that sixteen of the biggest ships in the world produce more pollution than ALL the cars in the world. Shipping is by far the biggest transport polluter in the world and the problem is with the heavy fuel oil the ships run on and the almost complete lack of regulations applied to the giant exhaust stacks of these ships. As we have seen with motor vehicles, oil and diesel produce oxides, but ships produce LARGE amounts of sulphur oxides (SOx) and nitrous oxides (NOx) when they are burned. This is the reason why the 16 largest ships in the world produce more of these oxide emissions than ALL of the cars in the world. Sulphur oxides and nitrous oxides are very harmful to the environment in many different ways, dramatically adversely effecting ecosystems as they can kill marine life and cause respiratory illnesses in humans. Airborne pollution from these giant diesel engines has been linked to sickness in coastal residents near busy shipping lanes. Up to

60,000 premature deaths a year worldwide are said to be as a result of particulate matter emissions from ocean-going ship engines. It is argued that a very simple way to solve this issue would be to switch shipping fuel to Liquid Natural Gas (LNG). It can be said that switching all maritime vessels to LNG as the main fuel would represent the single biggest technological advance in the shipping industry of the last 100 years.

Aircraft contrails

Having discussed land and sea transport it would be remiss to not mention air travel - that of global commercial and military aircraft - which of course, is another source of emissions and pollution. According to 'Environmental Protection UK' which is a national charity that provides expert policy analysis and advice on air quality, land quality, waste and noise and their effects on people and communities in terms of a wide range of issues, including public health, planning, transport, energy and climate, claimed that in 2011, approximately 200 million passengers passed through mainland UK airports. This was a return to growth, following a recent period of decline in passenger numbers and air transport movements between 2007 and 2010. Government forecasts predict that this will rise to 255 million in 2020 and 313 million in 2030. Aircraft are responsible for an increasing proportion of air pollutant emissions, both at local and global level. Aircraft engines generally combust fuel efficiently, and jet exhausts have very low smoke emissions. However, pollutant emissions from aircraft at ground level are increasing with aircraft movements. In addition, a large amount of air pollution around airports is also generated by surface traffic. The main pollutant around airports is nitrogen dioxide (NO2) and is formed by nitrogen oxide (NOx) emissions from surface traffic, aircraft and airport operations. Nitrogen oxides from high-altitude supersonic aircraft are thought to damage the stratospheric ozone layer, the protective layer that filters out harmful radiation from the Sun.

CONTRAILS IN AN OTHERWISE
CLOUDLESS SKY

I would suggest that CO_2 emissions from aircraft, although undesirable, are not, relatively speaking, a significant problem as they contribute approximately 3 per cent of the global total. The more concerning problem being water vapour aircraft emit, which shows up as condensation trails (contrails) behind high flying aeroplanes. Indeed a few years ago this was amply illustrated when sunbathing during a very sunny morning in June on a beach in Tenby, Wales. It was not long before the sky was covered in a tenuous high cloud covering due to various aircraft contrails converging. This had the effect of diminishing the heat from the Sun, putting an end to a glorious sunny and warm June morning thus making sunbathing quite uncomfortable. How many people are aware that water vapour (contrails) released at high altitude of greater than 30,000 feet has a bad effect on the atmosphere many times greater than it would if released in the lower atmosphere, as the vapour does not condense into clouds and rain in quite the same way, and as such collects more dust, et cetera – this effects the way the atmosphere behaves at those heights – could recent weather anomalies be attributed in some degree to this effect, and if so, will the situation worsen with more and more aircraft filling our skies?

Conclusions

THE INTRODUCTION OF ELECTRIC VEHICLES SHOULD BE DELAYED AND RE-ASSESSED AS THERE ARE MANY DIFFICULTIES TO BE RESOLVED. PROBLEMS SUCH AS THE

COST OF EACH VEHICLE, LACK OF INFRASTRUCTURE, LONG CHARGING TIMES AND THE LACK OF THE NECESSARY CHARGING POINTS. THESE DIFFICULTIES WILL CERTAINLY BE WORSENED BY THE SIGNIFICANT DEMAND FOR EXTRA ELECTRICITY AND THE LOSS OF BILLIONS IN FUEL AND VEHICLE EXCISE DUTY. THERE IS, OF COURSE, THE DILEMMA OF DISPOSING OF 40 MILLION FOSSIL-FUELLED VEHICLES, NOT TO OVERLOOK AT LEAST 160 MILLION TYRES.

UNFORTUNATELY GOVERNMENT STRATEGY FOR AN EVER INCREASING DEPENDANCY ON WIND AND SOLAR POWER GENERATION, COUPLED WITH THE REPLACING CONVENTIONAL BOILERS WITH HEAT PUMPS, WILL WITHOUT DOUBT, EXACERBATE THE ILL-CONSIDERED GOVERNMENT POLICIES.

ELECTRIC VEHICLES DO NOT HAVE THE SAME NEUTRAL GEAR AS COMBUSTION ENGINE VEHICLES. THEREFORE MANUFACTURERS ADVISE AGAINST TOWING ELECTRIC CARS. IN FACT DAMAGE CAN OCCUR IF ATTEMPTS ARE MADE TO TOW THE VEHICLE - ROADSIDE ASSISTANCE IN THE FORM OF HAVING THE CAR LOADED ONTO A TRUCK OR TRAILER WILL BE NECESSARY.

IT IS ACCEPTED THAT **IF** THE VARIOUS HURDLES CAN BE CLEARED, THEN THE FUTURE LOOKS HOPEFUL. BUT SADLY GOVERNMENT HAS LOST ITS WAY BEING DEFICIT IN TECHNOLOGICAL AND ENGINEERING KNOWLEDGE AND SKILLS. THIS LACK OF KNOWLEDGE, SKILL AND WISDOM UNLEASHES THE *BEAST* AND AS SUCH DARK CLOUDS ARE GATHERING ON THE HORIZON.

CHAPTER SEVEN

'Now I am become death, the destroyer of worlds'

J. Robert Oppenheimer (1904-1967)

Theoretical physicist and credited father of the atomic bomb.

NUCLEAR POWER

The safe use of nuclear fission energy demands the highest respect, discipline and professionalism in all its applications, both military and civilian. Indeed, in all military endeavours the possibility of any horrendous accident should be absolutely minimal, and every care should be taken especially when handling weapons of mass destruction. Needless to say the same principles should apply equally to Nuclear Fission Power Stations due to proximity of populations. A word of caution as some readers may find this chapter, depending on their sensibilities, uncomfortable and unsettling.

Fundamentally there are two types of nuclear power, namely fission and fusion. In fission, the nucleus of an atom splits into two or more smaller nuclei; in fusion, two or more nuclei combine. Each reaction releases great energy, which sadly and shamefully can be harnessed for destruction such as the atomic (fission) bombs dropped on Hiroshima and Nagasaki, Japan in 1945. The much more powerful Hydrogen bomb was detonated on November 11th, 1952, in Enewetak in the Marshall Islands. This thermonuclear device employed fusion reaction, and the bomb required the addition of a fission bomb to detonate - the fusion reaction is started with a fission reaction. But unlike the fission (atomic) bomb, the fusion (hydrogen) bomb derives its power from the fusing of nuclei of various hydrogen isotopes into helium nuclei. The world's first full-scale commercial nuclear power station Calder Hall, west Cumbria, was opened by the Queen in 1956. *When this first nuclear power station was opened the excited media announced that electricity would become so cheap, that no one would even bother metering it – how hollow those words sound these days.*

The site is now known as the Sellafield plant in west Cumbria. Workington became the first town in the world to receive heat, light, and power from atomic energy. During the working lifetime of Calder Hall it produced relatively little power and far less than the 1,198 MW output of the Sizewell B modern reactor in Suffolk. The Sellafield site incorporated Britain's first generation nuclear reactors and associated fuel re-processing facilities at Windscale, and as mentioned above the world's first nuclear power station to export electricity on a commercial scale to a public grid at Calder Hall. Sellafield is a large multi-function nuclear site close to Seascale on the coast of Cumbria, England; the UK's National Nuclear Laboratory has its Central Laboratory and headquarters at Sellafield. Built as a Royal Ordnance Factory during 1942, where in 1947 was used for the production of plutonium for nuclear weapons, being given the name 'Windscale Works'. Subsequent key developments included the building of Calder Hall nuclear power station. As of August 2020, activities at Sellafield include nuclear fuel reprocessing, nuclear waste storage and nuclear decommissioning. The licensed site covers an area of 265 hectares, and comprises more than 200 nuclear facilities and more than 1,000 buildings. It is Europe's largest nuclear site and has the most diverse range of nuclear facilities in the world situated on a single site. The Nuclear Decommissioning Authority (NDA) during 2019 said the news that Calder Hall is now empty of nuclear fuel for the first time, had brought the multibillion-pound clean-up operation a major step closer, and the full decommissioning cost is said to be more than an eye watering £70 billion. To put this astronomical figure into perspective, approximately 70 modern gas-fired power stations could be built *from the decommissioning costs alone*! In addition to Nuclear power stations incurring astronomical building costs of up to 30 times more than a natural gas-fired (CCGT) power station, nuclear power stations, as mentioned above, have horrendous de-commissioning, waste storage and disposal costs. There is not a single nuclear power plant anywhere on the planet that has been completely and safely decommissioned at the time of writing. Lord Marshall after retiring from the then Central Electricity Generating Board (CEGB), told a reporter that he always underestimated the costs deliberately to Parliament, as he knew they would never get built if he did! That was when we could design and build them in the UK – now that expertise has to be foreign-sourced and the building of the EDF Hinkley Point, in Somerset as a prime example.

SELLAFIELD WAS INTENDED TO PRODUCE PLUTONIUM FOR THE UK ATOMIC WEAPONS PROGRAMME

It is sobering to realise, that all the nuclear plants that have been shut in approximately the last 30 years, are still going through the stages of storage and processing of spent fuel and materials - this is conveniently ignored when considering nuclear generation - such is the madness. It is absolutely vital to appreciate the problems associated with hazardous materials (what a legacy for future generations), and the potential for an extremely nasty accident – the leakage of radio-active material at the very least. Nuclear power plants are designed, built and will be maintained by Humans - a species that does not have an impressive record that inspires confidence, being prone to arrogance, pride, greed, corruption, miscalculation, error, accident, acts of war and terrorism. It is also *instructive* to remember the quote attributed to John Glenn, an American astronaut, which offered the profound words, "As I hurtled through space, one thought kept crossing my mind - every part of this rocket was supplied by the lowest bidder."

Nuclear mishaps

Surely the expectation would be of the highest professionalism, discipline and safety when handling Weapons of Mass Destruction (WMD). But alas history has begged to differ as the following incidents will testify, and if such scenarios can arise with WMD, then what of Nuclear Power Stations close to populations, such as on the over-crowded mainland of the UK.

I wonder how many readers are aware of the near *thermonuclear disaster* off the Andalucian coast of Spain when on 17th January, 1966 a giant US Air Force giant B-52G bomber from the Seymour Johnson Air Force Base North Carolina, USA, collided with a KC-135 tanker aircraft while attempting to re-fuel at 9,450 metres (31,000 ft) causing a massive fireball which engulfed both aircraft – killing all four men on the tanker and three of the bomber's crew. Four other bomber crew members managed to eject before plunging into the sea where they were thankfully rescued by Spanish fishermen. Later the official investigation concluded that the B- 52G overran in manoeuvres to hook up with the trailing fuel boom and rammed the tanker. This aerial accident was horrendous in itself, but what made it a thousand times more terrifying was that the B-52G bomber was carrying four 1.5 megaton Mark 28 thermonuclear hydrogen bombs – each 70 times more powerful than the atomic bomb that destroyed Hiroshima in Japan in 1945. Imagine the holocaust had the bombs detonated causing thermonuclear explosions – it would have been truly horrendous. One bomb splashed into the Mediterranean Sea, another drifted down on its parachute landing in a dried-up river bed, with the other two splitting open on impact with the ground scattering plutonium and covering the tomato fields near the village of Palomares, Spain with a fine and deadly radioactive dust. Plutonium is a very toxic material and takes thousands of years to become safe, and that is why air monitors have been installed around Palomares since 1966. Despite numerous clean-up efforts, radioactive material continues to be found near the crash site, including two trenches filled with radioactively contaminated soil, which were discovered in 2008. Another concern is that plutonium decays into other radioactive components like americium, a gamma-emitter, which can harm people over large distances. It is now over half a century since this incident and it begs the question of what it will take to make the area safe again for not only agriculture, but for human habitation? The following example of another *nightmare* incident does not exactly inspire confidence. Unbelievably, it was only when an Albuquerque newspaper published an article based on military documents recovered through the Freedom of Information Act in 1986, the rest of the world learned of a terrifying accident just south of Kirtland Air Force Base on May 22, 1957. The city of Albuquerque and a good portion of the surrounding region were nearly obliterated by the accidental detonation of a 10-megaton hydrogen bomb, dropped by an American Bomber on the outskirts of Kirtland Air Force Base. The massive 10 megaton bomb was more than 600 times more powerful than the 'Little Boy' atomic bomb dropped on Hiroshima in World War 2. An investigation found the bomb had created a

12 feet deep and 25 feet wide crater - fortunately New Mexico was not blown into oblivion. The accident was caused by a crew member who lost balance when removing a locking pin, causing the hydrogen bomb to drop rapidly towards the desert below. Thankfully, no civilian was harmed when the bomb dropped.

I hope the above will not keep you awake at night! No doubt those who support Nuclear Fission will quote the favourite *'get-out-of-gaol-free'* line, that *'lessons will have been learnt'*. Unfortunately, the following incidents will sadly show that lessons have NOT been learnt:

- It is unnerving to realise that in 1965 a U.S. Navy Skyhawk jet bomber fell off the deck of a ship and sunk in 4,900 metres (16,000 ft) of water, 130 kilometres (80 miles) off the coast of Japan - the aircraft took a one-megaton hydrogen bomb along with it to Davy Jones's Locker.

- At a uranium reprocessing plant at Tomsk, Siberia, Russia during April, 1993, a tank exploded sending a cloud of radioactive particles into the atmosphere.

- How many readers are aware that in 1957 at Windscale (now renamed Sellafield), Cumbria, England, fire destroyed the core of a reactor, releasing large quantities of radioactive fumes into the atmosphere.

- With all the reassurances of the nuclear industry it is shocking that 380 million litres (100 million gallons) of radioactive water, containing uranium, leaked from a pond into the river Rio Purco, at Church Rock, New Mexico, USA in July 1979. The consequence of this was to cause the water to become over 6,500 times as radioactive as safety standards allow for drinking water.

- Scotland's Dounreay nuclear power station is expected to take until 2036 to be dismantled at an estimated cost of 3 billion pounds. It is an extremely complicated process. How many people are aware of its poisonous past? During the 1960's and 1970's radioactive material found its way into the sea! Proponents of nuclear power will argue that was a long time ago and things have changed dramatically since then. Really! I beg to differ, please read on.

- During an incident in 2004, thousands of gallons of radioactive material leaked unnoticed from Sellafield's Thorp reprocessing plant. It was said to be the most serious leak anywhere in the world in 2004. The problem was blamed on a faulty pipe which had suffered metal fatigue. There is always something or someone to blame.

If highly trained and disciplined military personnel cannot be trusted with weapons of mass destruction such as thermonuclear bombs, which by their very nature, demand extremely high standards of care and attention, then what of nuclear power stations placed on a relatively small island such as the UK. It is not rocket science to realise that it is just not worth the risk of building extremely expensive, and potentially harmful nuclear (fission) power station in the over-populated UK, when considering there are other viable generation options to consider. It takes about 10 years to complete a nuclear station, and costs up to 30 times more than a natural gas-fired power station; they are subject to delay and always overrun on costs, the EDF Hinkley Point Nuclear Fission plant being a classic case. We should be very reluctant to even contemplate the nuclear fission option in the provision of electrical energy. In fact we should not consider nuclear fission power as an alternative to anything unless there is a 100 per cent assurance to its safe use – and this is fundamentally an impossible dream. It is more than evident that all systems designed and built by humans are prone to error or failure - we are all fallible. Apart from horrendous cost and waste disposal, it is possible design and human failings that are other reasons to reject Nuclear (fission) Power Stations – and the more of them, big or small, the greater the risk.

The Three Mile Island disaster (1979) happened for its own *specific* reasons - such as the combination of mechanical and electrical failure as well as operator error - causing a pressurised water reactor to leak radioactive matter.

The Chernobyl disaster (1986) occurred because of its own *specific* reasons - overheating causing an explosive leak of clouds of radioactive material, resulting in large scale evacuations from Chernobyl and Pripyat – and the next significant Nuclear Power Station disaster will occur due to its own *specific* reasons! It is important to realise that almost 35 years after this nuclear disaster, crops near Chernobyl are still contaminated. Barley, wheat and rye nearly 30 miles away are still high in radioactive chemicals such as strontium 90 and caesium 137.

What are we to make of the 2011 disaster at the Fukushima nuclear power plant near Iwaki, Japan when three nuclear reactors went into meltdown! This was as a result of a 9.0 magnitude earthquake, causing a tsunami that engulfed the power station. Now who would have dreamt of, or even considered, building a nuclear power station in an area prone to earthquakes? Especially directly on a coastline that faces out to the Pacific Ocean – an ocean that is no stranger to the destructive forces a tsunami can bring - an *alien* might question whether Humans are the most intelligent and sensible life forms on the planet! What will it take to learn the lesson, when more than 1,600 people have died from health complications brought on by this particular nuclear disaster in 2011 – since then, and at the time of writing, the company that owns the plant, Tokyo Electric Power (Tepco), has struggled to bring the plant under control, with hundreds of litres of radioactive water flowing into the Pacific ocean. The clean-up effort is set to cost more than £11 billion - excluding the compensation still owed to thousands of families.

The lesson has certainly not been learnt by the UK government with the proposed new EDF nuclear power station, Hinkley Point C (3.2 GW capacity with two reactors), at an estimated cost (at time of writing) of more than £20 billion. The power station is to be situated in Somerset on the Severn Estuary, approximately five miles from Bridgwater, 15 miles from Minehead in the west and roughly six miles from Burnham-on-Sea. It should be noted that Hinkley Point nuclear power station is owned by French energy firm EDF ENERGY, and the China General Nuclear Power Group (CGN), has partnered with EDF to help fund a third of the £20 billion cost of the nuclear power station. I truly find such strategy unbelievable when involving foreign countries especially when the UK has many decades worth of natural gas to power efficient modern gas-fired power stations that are much *safer, cheaper,* and *quicker* to build, whilst offering a 60 per cent reduction in atmospheric emissions than their coal- fired cousins. Fracking in the UK would turn the economy around the same as it has in the USA. There is an existing nuclear power station at Hinkley Point, namely Hinkley Point B power station. This was the first Advanced Gas-cooled Reactor to generate electricity to the Grid in the UK, and consisted of two Advanced Gas-cooled Reactors supplying a total of 955 MW to the National Grid. Construction started in 1967 and generation started in 1976 – a total of 9 years from start of construction to actual generation, with an estimated decommissioning date for 2023. During March 2009, Hinkley Point C was officially nominated as a potential site for a new nuclear power station by the Government, with construction, unbelievably, on low-lying coast in the Severn estuary – has the recent lesson of the Japanese Fukushima nuclear plant disaster fallen on deaf ears? Government, it would appear, is blind to history - aided and abetted by its technological and engineering lack of common sense. On 30 January 1607, the Severn Estuary was devastated by a great flood which proved very extensive on both sides of the Bristol Channel. On the Welsh side, flooding was devastating from Laugharne in Carmarthenshire to above Chepstow in Monmouthshire. The coasts of Devon and in particular, the Somerset Levels as far inland as Glastonbury Tor, 23 kilometres (14 miles) from the coast, were also affected. The sea wall at Burnham-on-Sea collapsed, and the water flowed over the low-lying levels and moors. If this calamity were to happen again, what hope for a low lying coastal nuclear power station such as Hinkley Point? Armageddon is the word that quickly springs to mind. The inundation of a nuclear power station and a possible meltdown would make the 1607 disaster pale into insignificance. The 1607 flood affected thirty villages in Somerset, including Brean which was 'swallowed up' and where seven out of the nine houses were destroyed with 26 of the inhabitants dying – across the area an estimated 2,000 or more people drowned, with houses and villages swept away - 200 square miles (51,800 ha) of farmland inundated and livestock destroyed. The reasons for the great flood of 1607 has been attributed to massive undersea boulders being displaced, resulting in rock erosion and triggering high wave velocities throughout the Severn Estuary. The other possible cause being a storm surge - similar to the 1953 North Sea Flood, where high tides and a storm surge resulted in floods in East Anglia, Canvey Island, the Netherlands, Belgium, and Scotland. The Bristol Channel floods drowned many people and destroyed a large amount of farmland and livestock; recent research after investigations by Professor Simon Haslett of Bath Spa University and Australian geologist Ted Bryant of the University of Wollongong, has suggested that the flooding may have been caused by a tsunami. There has been flooding around the Bristol Channel numerous times in the past century. During the winter months of 2013-14 large parts of the Somerset Levels were under water, such that villages were isolated, homes evacuated, and the farming community was in disarray as the bad weather was relentless with incessant rain. A storm surge which coincides with a spring tide is rare, but when it does occur can have devastating consequences. As sea levels rise, so the risk increases, and what is now rare may become more common within a century.

Power stations need access to water for cooling, and so they are built near to water. It is claimed the Hinkley site has a low flood risk, but with sea level rise the flooding risk will become more threatening and a much higher level of risk. Such storm surges coupled with spring tides are of course rare, but is the combined risk coupled with sea level rise now acceptable, there is too much at stake! A serious meltdown and explosion at Hinkley would immediately make Bristol, Gloucester, Newport and Cardiff uninhabitable, and depending on wind direction could cause *nightmarish* problems elsewhere. Is it worth the risk? It is wishful thinking to consider that we are all safe from a disaster at a nuclear power station – and an ignorance of history means we will bound to make the same mistakes - there will be another mishap - it could be due to nature, human, equipment failure, a combination of events, or maybe an act of terrorism - the terrifying thought is when and how widespread and disastrous will it be? Supporters of fission plants will argue that more deaths have occurred within the conventional power generating industry than at nuclear plants -

but the power industry has been going since the 1880's, whilst the nuclear industry has only been running since the opening of Calder Hall by Queen Elizabeth 2 in 1956. It is somewhat pointless in comparing the accident statistics of the two industries, as it will only take one major disaster at a nuclear plant in the UK, with the potential of killing and injuring hundreds of thousands, or maybe millions of people and contaminating acres and acres of land with radiation, to make a complete nonsense of any safety statistics. Do we really need to take this risk on our tiny and over-crowded island, when there are so many other options? In a sense the 'real' question is not whether or not we need nuclear fission power, but whether we can afford a disastrous mishap on our tiny and over-crowded island. Governments controlling large land masses in the Americas or Asia such as continental America or Russia may be willing to take the risk - they have the space - and have indeed escaped an actual Armageddon scenario so far.

But, what of other existing nuclear plants, especially those across the English Channel in France - after all, France is not that far away as the crow flies. Power generation in France mainly comes from nuclear power, which accounted for over 70 per cent of total production in 2018, while renewable and fossil-fuels accounted for 21 per cent and 7 per cent, respectively. France has the largest share of nuclear electricity in the world, and the nuclear power sector is almost entirely owned by the French government. The concern is whether the French have just been lucky so far, not to have experienced a catastrophe such as Chernobyl or Fukushima. It is irresponsible to minimise the terrible incidents of Chernobyl and Fukushima - the first incident was caused by human failure, and the second, by forces of nature – although it can be argued at Fukushima the prime cause was human failing again, simply by building a Nuclear Power Station in the wrong location. How many readers are aware of how close France has come to its own nuclear disaster? Floods, earthquakes, heat waves, and exceptional cold have all caused a number of problems for the French nuclear industry, mostly in their effect on crucial cooling systems. Many of the threats to safety involved faulty mechanisms
- in the 1990s the problem was sump-clogging which could seriously impair the emergency cooling system. It should be noted that in a nuclear power Station's reactor housing, the role of the sump is to collect any overflow of primary loop coolant. The monitoring and pumping of the sump is an important part of the reactor's safety system.

A violent storm (Cyclone Martin) hit France in on the night of 27th - 28th December 1999, and wind speeds reached around 200 km/h (120 mph) in French département of Charente-Maritime. In south western France, flooding occurred at the EDF Blayais nuclear plant in Gironde and nearly caused a major incident. The flood defences were breached and about 100 million litres of water poured into the plant. Power from the external grid failed. Two of the four reactors went into shut-down where key safety equipment failed as a result of flooding. Finally, one half of the emergency circuit cooling system failed, though it held out long enough to bring the situation under control. Local officials admitted at the time that they were close to a fuel meltdown, and the similarities to Fukushima are frightening.

Not wishing to be alarmist, and with the greatest respect, but to reiterate that the Hinkley Point Nuclear Power Station is being built by EDF on the coast in Somerset which experienced severe flooding in the 1600s. No doubt, and we pray, that those responsible for constructing this power station have taken all possible threats into account against any possible nuclear incident.

During 2008, a series of nuclear accidents sent shock waves throughout France, severely challenging the façade of competence and safety that France had created.

- In March 1980, an accident at EDF's Saint-Laurent nuclear reactor in central France caused two fuel rods to melt, seriously damaging the reactor and causing the then most serious accident in France's nuclear history.

- The Tricastin nuclear power station in southern France malfunctioned in July, 2008, causing 30,000 litres of a solution containing 12 per cent enriched uranium overflowing from a reservoir into the nearby Faffiere and Lauzon rivers. This raised the concentration of uranium in the two rivers a thousand times. That was only the first of a series of 9 blunders and leaks in France's nuclear plants in 2008.

- In July 2008, thirty cubic metres of liquid containing natural uranium was accidentally poured on the ground and into a river, at Areva's Socatri site in south eastern France. Pure uranium is not as dangerous as enriched uranium, but France's ASN nuclear watchdog rebuked Areva for mishandling the accident.

- During November 2009, a fuel assembly rod became stuck in the pressure vessel at EDF's Tricastin plant in southeast France, raising the risk of an accident. A similar incident took place in September 2008 in the same reactor during refuelling operations.

- In June 2011, a minor and fairly common incident involved internal leakage at EDF's Paluel 3 nuclear reactor.

What should be of great concern is that the average age of EDF's French 58 reactors is 35 years, and most started up late 1970s to early 1980s. The cost of ensuring the highest levels of safety is enormous. France and other European nations are carrying out stress tests to assess the safety of their reactors, and Germany's coalition government has announced that all its nuclear power plants will be phased out by 2022. This decision makes Germany the biggest industrial power to announce plans to give up nuclear fission generation.

It would appear that the French have been lucky not to have - as yet - experienced a catastrophe similar to Chernobyl or Fukushima. At the end of the day common sense dictates that we are ALL better placed taking our chances with emissions from combined cycle gas turbine stations, than with the nuclear fission option - remember, as pointed out above, it will only take one major disaster to prove the point. They said the Titanic was unsinkable, but human misjudgement and icebergs had other ideas! Could the Tacoma road bridge collapse in the United States have been prevented, or the collapse of the Tay Bridge in Scotland with its unfortunate passenger train, and what of the mishaps attributable to nuclear submarines? Then we have the space shuttle disasters and the loss of two Mars Polar Landers' during 1999 at a cost of 250 million dollars. Indeed, one of the Mars Polar Landers failed as a result of - wait for it – a number of people working in Metric whilst others were using Imperial measurements, with the consequence of loss of navigation in the space craft and 80 million dollars down the drain. They say life is stranger than fiction – and events of this nature certainly prove it.

It is instructive to note when assessing risks and taking air travel as an example, air accidents are often caused by a chain of events, each relatively harmless in itself, but catastrophic when linked together. Reasons for plane crashes as sourced from the Aircraft Crashes Record Office are as follows, Human error: 68 per cent, Technical failure: 20 per cent, Weather: 6 per cent, Sabotage: 3 per cent and Freak causes: 3 per cent. The miraculous escape of flight BA038, a Boeing 777, carrying 150 passengers and crew, at Heathrow airport on Thursday 17th, January, 2008, should concentrate minds to the skills and also, unfortunately, the fallibility of Humans and their activities. Experts claimed that the chances of a double engine failure occurring on a modern aircraft were 'a million to one' – and although a catastrophe was avoided by the skill of the crew, it should be fully noted that both engines did indeed fail simultaneously and at a very critical point in the flight! It can be argued that it was lucky that the plane failed just at that particular moment in the flight. Slightly earlier the aircraft may not have made it to the airport, coming down on a very congested urban area. The weather was good and the aircraft was not fully loaded; the 209ft-long plane can carry between 305 and 440 passengers at a cruising speed of 615 mph with a range of up to 8,300 miles. The Boeing 777 was launched in June, 1995 and is considered an extremely reliable aircraft with an almost impeccable safety record; the plane is powered by two Rolls Royce engines and should still fly if one fails – its landing gear is the largest of any commercial aircraft. I quote the aircraft incident as it is a timely reminder to the advice of 'experts' and what could have been a major disaster killing hundreds of people – an incident at a nuclear fission power station has the potential to poison by radiation and/or kill thousands, if not millions.

It is sobering to recall the cautionary tale that Robert Gates, the American Defence Secretary can offer about Zbigniew Brzezinski (when he was President Carter's national security adviser) being woken up with the news that 200 Soviet missiles were on their way to America. Brzezinski demanded confirmation before alerting the president. Two minutes later, he was informed that the radar now showed 2,000 missiles. Just before he woke Carter, to tell him he had perhaps two minutes to launch a retaliatory strike, he was telephoned again to be told there had been an error, and that someone had put an exercise tape into the computers by mistake. Therefore, you will no doubt understand my fears about not having too much faith in human reassurances and guarantees. With hindsight most, if not all past disasters could have been avoided, so we must not fall into the trap and arrogantly think that we now know it all – even with the best of training and the latest technology we do not (nor ever will) have all the answers. Defenders of nuclear fission plants will point to a World Health Organisation (WHO) survey of the reactor incident at Chernobyl in 1986, which claims that the incident killed only 75 people, most of them either operating the plant at the time or rescue workers at the scene - although it is recognised the zone around the plant, evacuated nearly 35 years ago, has now become a thriving nature reserve. Now whether the WHO figure and observations are correct or not it is difficult to determine the truth, although I suspect a lot more suffered, if not fatally, from numerous associated radiation sicknesses. Nevertheless, there was an *incident* and this is the *important point* - next time it may be much, much worse. It is wise to keep in mind a claim for the 1979 accident at Three Mile Island, that the number of deaths reported seems to have been zero - what of associated radiation illnesses - were there none? An incident did happen and that is fact. It is argued that a breach in a large dam could easily kill tens of thousands and destroy hundreds of properties downstream, with far greater numbers in the case of the giant Three Gorges project in central

China. Research from The Paul Scherrer Institute, a Swiss government physics research laboratory, indicates that nuclear fission power has been responsible for a tiny fraction of fatalities, a fortieth, of that for renewable hydroelectric power. But what are these arguments trying to say, as it will only take a single major nuclear fission power station disaster to make a mockery of all these claims - surely this a chance not worth taking?

Supporters of nuclear energy claim there will be a significant reduction in atmospheric emissions if the UK changed to nuclear power – a claim similar to that of the supporters of wind-generated electricity. But this is a somewhat hollow and disingenuous argument - as China is a large producer of industrial energy and has a seven-year plan to build over 500 coal-fired power stations. Consider all those potential emissions - the UK is responsible for less than 2 per cent of the world's emissions - and there are also huge emissions from India, Mexico, Brazil, et cetera. Those readers that cannot agree with all, or some of the above observations, and consider the nuclear fission option is worthy of consideration, then they will still have to address the following: The prohibitive cost of nuclear fission energy as the cost is in the billions of pounds, and I doubt if there is anybody who will actually know the true costs - Guy Dauncey with Patrick Mazza in their book titled, 'Stormy Weather', state that in the U.S., the nuclear industry has already received $145 billion in subsidies, compared to $5 billion for solar and wind energy. In the United Kingdom the Nuclear Decommissioning Authority (NDA) estimates for the decommissioning of the country's nuclear power sites would be of the order of £70 billion. How many conventional power stations could be built for this amount of money, and how much environmentally-friendly energy research and development would this had funded? It really does make you think – what if this money could be spent on developing tidal generated electrical energy, energy saving devices and home insulation – a very worthwhile goal as it would lead to a significant leap in energy saving – think of that in environmental terms.

To summarise:

- Nuclear power stations are indeed highly expensive to build, costing in the region of about £23 billion for a 3,200 megawatt reactor, whilst the cost of a similar capacity natural gas-fired (CCGT) plant would be in the order of about £1.5 billion.

- The long time-scales involved in decommissioning old nuclear plants.

- The problem of what to do with nuclear waste! Nuclear waste stays highly radioactive for many thousands of years: a very real problem and future generations will not thank us for this.

- The possible use of stolen plutonium for nuclear weapons.

- Acts of terrorism: it would be very foolish and irresponsible to consider a nuclear power station 100 per cent absolutely safe from acts of terrorism.

During May, 2006 it was reported in the media that there have been 57 nuclear safety incidents around Britain since 1997 - the incidents ranging from equipment failure, radiation leaks to a fire and ground water contamination. An estimate (2013) by the United Kingdom's Nuclear Decommissioning Authority predicted costs of at least £100 billion to decommission the 19 existing United Kingdom nuclear sites. It should be noted that Germany has set aside £33 billion to decommission 17 nuclear reactors, and the UK Nuclear Decommissioning Authority estimates that the clean-up of UK's 17 nuclear sites will cost between £95–218 billion over the next 120 years. But who truly knows the time-scale, and total cost to the complete environmental cleaning up of nuclear plants?

Rolls-Royce small nuclear plants

At the time of writing Rolls-Royce, the British jet engine maker, said that it was forming a business to build a series of smaller, cheaper nuclear reactors as the UK looks for ways to cut carbon emissions and to reduce the costs of nuclear energy. The small modular reactor (SMR) group includes the National Nuclear Laboratory and the building company Laing O'Rourke, which has already received £18 million to begin the design effort for the SMR concept. It is proposed that the reactor would cover about two soccer fields, or about one-tenth the acreage of a conventional nuclear power station. The proposed plants would generate less power of approximately one-seventh the output of the giant nuclear (EDF) installation being built at Hinkley Point in southwest England. It was reported that Rolls-Royce hoped to reduce construction costs to around 2 billion pounds each, compared with an estimated £22 billion to £23 billion for the Hinkley Point power station. It is claimed that some of the savings would come from building a large number of plants and making modules in factories that can then be assembled at sites. The target is to build 16 small modular reactors, powering around one million homes - the Government will contribute a grant of £210 million to develop the plants. Both Hinkley Point and the concept of small modular fission reactors are considered hideous and reckless by the author for the many reason given in this chapter. Indeed, spreading risks around the country with smaller plants truly beggars belief. I also wonder if Government are playing on the good name of Rolls-Royce to sell the concept to the public – a question of smoke and mirrors – such as the misuse of the term wind turbine, something which it is not!

Nuclear (fusion) power stations

It should be clearly understood that nuclear fusion reactors do not use a *radioactive source* to generate power as they run on hydrogen. Two hydrogen atoms are fused together to form helium atoms, which produces neutrons and vast amounts of energy. Basically you put hydrogen in and you get helium and energy out - there are no spent fuel rods to contend with, and more importantly no risk of meltdown. Nuclear fission depends on the process of splitting an atom, as mentioned earlier. Without dispute nuclear fusion is the HOLY GRAIL and the most 'environmentally friendly' terrestrial means for generating electrical energy on the planet, offering safe and limitless energy. Remember fusion occurs constantly on the Sun, which produces energy via the nuclear fusion of hydrogen into helium. Nuclear fusion is a safe nuclear reaction because the meltdown of the nuclear reactor is physically ***impossible*** with no risk of nuclear disasters similar to the incidents that took place in Chernobyl and Fukushima.

Scientists have pursued the harnessing of nuclear fusion for almost three-quarters of a century, with the UK Atomic Energy Authority patenting a fusion reactor in 1946, but alas without reliable success. Indeed, nuclear fusion has shown great promise for decades but is yet to be viable on a commercial scale due to the fact that maintaining a fusion reaction requires more power than it generates. Although, recent advances in the quest for fusion power have reignited hopes that it can be made feasible. The Massachusetts Institute of Technology (MIT) claim the dream of nuclear fusion is on the brink of being realised, and according to a major new U.S. initiative claim they have the science, speed and scale to put carbon-free fusion power on the grid in 15 years. The project, with collaboration between scientists at MIT and a private company, will take a radically different approach to other efforts to transform fusion from an expensive science experiment into a viable commercial energy source.

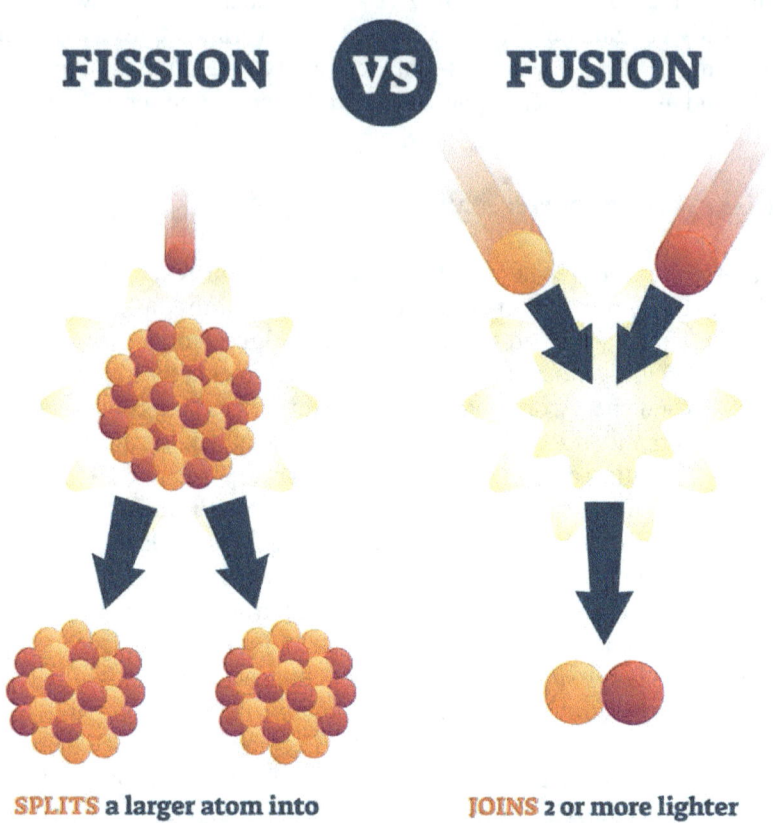

FISSION VS FUSION

SPLITS a larger atom into
2 or more smaller ones

JOINS 2 or more lighter
atoms into a larger one

Nearer to home, five sites in England and Scotland are in the final running for the UK's prototype fusion energy plants. The Government is backing plans for the Spherical Tokamak for Energy Production (Step) with a final decision on its location expected at the end of 2022 – but why not now! Nevertheless, let us be thankful for small mercies, as it would appear that there are some far-sighted and clear heads amongst the dunderheads in Westminster after all, thank heaven. The UK Atomic Energy Authority (UKAEA) said that the site would create thousands of jobs and aim to generate a *near-limitless* source of low-carbon energy, and the plant should be operational by the early 2040s. The five shortlisted sites are: Ardeer, North Ayrshire, Goole, East Riding of Yorkshire, Moorside, Cumbria, Ratcliffe-on-Soar, Nottinghamshire, and Severn Edge, Gloucestershire.

It was welcome news to read during February, 2022 that scientists have made a major breakthrough towards practical nuclear fusion. The Joint European Torus (JET) reactor laboratory in Oxford, has exceeded its own world record regarding the amount of energy it can extract by squeezing together two forms of hydrogen - managing to produced 59 megajoules of energy over five seconds - enough energy to boil the water in approximately 60 kettles full of water. The energy produced being more than double the previous record set in 1997, according to UKAEA. This latest success to a safe, efficient and low carbon means of addressing the effects of atmospheric pollution inspires confidence and should persuade politicians to increase significantly money available for the relevant research. Indeed, the success justifies the design choices for a bigger fusion reactor now being constructed in France.

'Without hesitation, I would like nuclear fusion to become a practical power source. It would provide an inexhaustible supply of energy, without pollution or global warming'

Stephen William Hawking (1942 - 2018)

The English theoretical physicist, cosmologist, and author.

The Director of research at the Centre for Theoretical Cosmology at the University of Cambridge

CHAPTER EIGHT

'If you can keep your head when all about you are losing theirs
and blaming it on you, If you can trust yourself when all men
doubt you, But make allowance for their doubting too; If you can
wait and not be tired by waiting, Or being lied about, don't deal in
lies, Or being hated, don't give way to hating, And yet don't look
too good, nor talk too wise'

Rudyard Kipling (1865-1936)

ALTERNATIVE ENERGY

Let us be brutally honest and admit that our myopic and inept leaders have totally failed us all. To be sure, it can be argued the lack of a secure and affordable electricity network that we are compelled to live with can be traced back to the Labour Party Energy Secretary, Ed Miliband and his Climate Change Act of 2008, which legally and disastrously committed the UK to cut carbon emissions by 80 per cent by 2050. The Liberal democrats can also hang their heads in shame, especially when the then Energy Secretary Ed Davey in 2013 boasted that the £400 million Pen-y-Cymoedd wind farm in south Wales would attract billions of investment to the UK and it would create hundreds of jobs – it is derisible that he claimed it was vital to secure our energy supplies for the future – I wonder what his thoughts are these days regarding the current energy crisis? The Conservatives have absolutely nothing to crow about either, as in 2019 Prime Minister Theresa May ridiculously increased the target to net-zero, and of course, Boris Johnson has eagerly embraced this absurd target, pursuing it like a madman – that is, at the time of writing – there may be a different Prime Minister by the time this book is published? It truly beggars belief that our leaders have placed Britain in such a precarious position as to be dependent on energy imports from foreign countries with gas imported from Qatar, U.S., Norway and Russia, especially when the UK is sitting on huge reserves of gas and oil. It should not go unnoticed that Britain is at the western end of the European gas grid, which is fed largely by Russia. Thus if Russia, for whatever reason, decides to restrict supply, there are no prizes for guessing who will suffer most. It should be noted that Russia was the largest exporter of natural gas to the EU in 2019 and 2020, which accounted for more than 40 per cent of EU imports. Why are we importing coal, of all things, from U.S., Russia and Venezuela, and there is something 'Alice in Wonderland' about importing dirty coal in diesel engine ships over four thousand miles from Venezuela – surely even the 'Mad Hatter' understands what is meant by a carbon footprint? We also import oil from, Nigeria, Saudi Arabia, U.S., Norway, Netherlands and Russia. It seems incredible that just twenty years ago the country was self-sufficient in oil - now our biggest importer is Norway. Just to rub salt into the wound we import electricity from France, Belgium and The Netherlands, and unbelievably there are plans for additional Interconnectors, as disclosed in this chapter, so much for energy security. The different means by which electrical energy is currently produced all have varying degrees of unacceptability, and all types of electricity generating power plants have an effect on the environment, one way or another - some power plants such as coal obviously have larger effects than others such as hydro – the 'trick' is to try and minimise these environmental impacts. The effect of power stations can be divided into two scenarios such as, fuel sourcing, atmospheric and localised pollution, and secondly construction, such as manufacturing, installation, decommissioning and disposal. Coal-fired power stations release significant amounts of carbon dioxide (CO_2) and sulphur dioxide (SO_2) into the atmosphere which, as we all know, are detrimental to the environment. Compounding this effect the coal mining industry has a long history of negative environmental impacts on local ecosystems, health impacts on local communities and workers, and contributes negatively to the global environment such as diminished air quality. It is for these very reasons, that coal has been one of the first fossil fuels to be phased out of in the UK. Sadly many major coal producing countries, such as China, India and Australia, have not yet reached peak production, with output increases replacing the falls in Europe and USA. Nuclear fission power plants are a real headache when it comes to decommissioning and the disposal of deadly nuclear waste.

The million dollar question has to be, where do we go from here - bearing in mind that privatisation of the power industry and the coercion for customers to have the innocuous Smart Meter fitted is an affront to consumer intelligence. Wind farms and Solar Parks by their very nature, will not offer security of supply or a significant level of supply for the National Grid as we have discovered in earlier chapters. It has also been shown that wind and in particular solar (roof-fitted) have their use on a smaller scale. Government is blind to the opportunities offered by the potential of marine energy being *foolishly* intent on nuclear fission, wind and solar. Government would have us believe that the only feasible way forward is predominantly large scale wind and solar farms, and nuclear fission, aided by biomass, and interconnectors. This is utter nonsense as there are numerous alternatives and options of varying degrees of acceptability as we now shall see.

Interconnectors

Fundamentally UK interconnectors are submarine power cables, offering a means for electricity to flow between separate alternating current (AC) networks, or to link synchronous Grids. They permit the trading of electricity between countries such that a country that generates more energy than it requires for its own activities can sell surplus energy to a neighbouring country. Interconnectors can also take the form of underground or overhead power lines. It is argued that interconnectors enhance the security of the energy supply and to manage peak demand. It is also claimed that the cross-border access to producers and consumers of electricity, increases the competition in energy markets, and they help integrate more electricity generated from renewable sources thereby reducing the use of fossil fuel power plants and thus CO_2 emissions. Interconnectors aid adaptation to changing demand patterns such as the uptake of electric vehicles. But I would challenge all these claims as the UK has the ability to generate all the power it requires and unless these interconnectors are intended to *sell* UK electricity to other countries at a profit, why do we need them? It is stretching credibility to claim interconnectors assist the uptake of electric vehicles – what next - interconnectors will help you sleep at night! Recognise also that from a security aspect we should not become dependent on power from foreign sources. The threat by the EDF, the French-owned electricity company, to turn off the power to Jersey (May, 2021), should act as a timely reminder not to be dependent on other countries for UK electricity. The fire during September, 2021, causing the French Interconnector to close down should be seen as the '*writing on the wall*'. Additionally, electricity does not take kindly to long power routes due to the inevitable power losses in the cables. Therefore before looking at a number of existing and proposed interconnectors it will be helpful to remind ourselves of losses in conductors and how modern power distribution came about.

When planning and constructing electrical circuits and distribution networks, it is very important to be aware of resistance and power loss in conductors and components. From the simple equation Power equals Voltage square divided by Resistance, it is easily seen that for a given voltage, if the resistance of an electrical circuit increases, then the power available diminishes; this is commonly known as *power loss*. If the voltage is increased for a given resistance, then power is not lost, hence the use of Very High Voltages in the long routes of the National Grid. Historically the early power stations were unable to distribute direct current (DC) electricity over any appreciable distance due to the loss of energy in the line conductors in the form of heat. Due to this power loss, useful current over a large distance would require metallic line conductors of large diameters which are very expensive, making transmission over large distances uneconomical. A major disadvantage with the early central generating stations was that as the system expanded, and the wires extended further and further, the losses from the wires became more and more expensive. Due to this cost the use of heavier and heavier wire was not the answer. But an answer was found in the use of alternating current (AC) electricity, and with the advent of the transformer (1831) this new device could raise or lower the voltage as required - an electric wire carrying a given amount of power, the higher the voltage, the lower the current and therefore the lower the losses in the wire. Therefore an AC generator at a power station could have its voltage output stepped up by a transformer - this 'stepped up' voltage could then be transmitted over the network wires with minimal power losses - at the far end the voltage could then be stepped down, for customer usage. The AC system, due to its minimal power losses, could cover a substantially larger area than a DC system. It also meant that electricity could be generated at remote sites where water was freely available and could be employed to produce what is known as hydroelectricity. In the early days most electricity was produced this way and in many parts of the United States electricity was commonly known as 'hydro'. The provision of high-voltage electricity became known as TRANSMISSION. To distinguish from the low-voltage electricity that was fed to customers the low-voltage was called DISTRIBUTION.

George Westinghouse of the United States was a leading pioneer in the use of AC and during the 1880s there was fierce competition between the DC and AC systems, each having its various merits. The DC system was relatively simple and not so costly for customers living close together in city centres. Whereas AC was more economical for customers spread over a large area in the suburbs and rural areas. Additionally, in those early days DC could run motors whereas AC could not. It was during the 1880's when 'three-phase' was introduced, which reduced the number and size of wires such that the AC system had another cost cutting boost. With the advent of the transformer offering the ability to convey electrical energy over large distances, this resulted in large central AC power stations becoming economically viable. The growing success of the AC system meant that hydroelectricity became very popular, especially in the United States, due to the fact that there were plenty of water resources. One of the most famous engineering projects in the United States was the construction of a large hydroelectric dam across the Colorado River, known as the Hoover (Boulder) Dam. Countries such as the United Kingdom did not have the same water resources as the United States, but luckily they had plenty of fossil fuel in the form of coal. Thus it was not surprising that Britain's early power stations were fired by coal. Unfortunately the burning of coal covered the local neighbourhood with smoke and soot and then there was the problem of getting rid of the ash. Additionally, the noise and vibration of the early steam engines shook the surrounding buildings; living near one of these power stations was not an endearing experience. The early steam engines were not very efficient in extracting the chemical energy out of the coal and had 'fuel efficiencies' of less than 10 per cent, that is, over 90 per cent of the energy in the coal was wasted! Indeed, the 'thermal efficiency' of steam locomotives and steam engines over a century ago was no more than 2 per cent, whereas the 'thermal efficiency' of a *steam turbine* is about 40 per cent. The power systems today use AC. The mains electrical supply in the UK is generated at a frequency of 50 Hertz (Hz) and is delivered to houses at 230 Volts (V). Noting that typical utility-scale (Grid) power plants generate AC electricity, and most electrical loads run on AC power. As such the majority of transmission lines carrying power around the world are of the AC type. But there are instances when High Voltage Direct Current (HVDC) transmission systems offer significant benefits. HVDC technology is used to transmit electricity over long or very long distance by overhead transmission lines or submarine cables, as it then becomes economically attractive over conventional AC transmission lines. If the transmission line route is longer than about 300 miles, DC is a better option, because AC lines have more line losses than DC for bulk power transfer. A disadvantage though of using HVDC is the investment costs, as HVDC converter stations are more costly than for high voltage AC sub- stations. Converter stations are needed to take power from an AC network then convert it to DC at the converter station; it is then transmitted to the receiving end by a HVDC submarine cable, underground cable, or overhead line, and then converted back to AC in another converter station before feeding into the receiving end AC network system. The initial cost can be prohibitive and as such is an important variable in the equation. Therefore HVDC transmission is usually restricted to very long distances. But over their lifetime, it is claimed that long-distance transmission lines can save operators money because of their efficiency, coupled with the fact that HVDC also allows power transmission between unsynchronized AC transmission systems - the power flow through an HVDC link can be controlled independently of the phase angle between source and load, it can stabilize a network against disturbances due to rapid changes in power.

The North Sea link

The North Sea Link (NSL) is a Norwegian–UK interconnector comprising a 1,400 MW subsea HVDC cable under construction between Norway and the United Kingdom. The construction was completed in June 2021 and became operational in October 2021. It is currently the longest subsea interconnector in the world. It is claimed the link will connect the Nordic and British markets directly for the first time, providing significant benefits for both countries – it will provide additional transmission capacity for electricity to be traded more efficiently, and will contribute to downward pressure on electricity prices when demand is high on one side of the cable. Really! When dear reader did your electricity bill show a fall in price and why would Norway buy electricity from the UK when it has an abundance of hydropower? Unlike the United Kingdom, generation in Norway is predominantly from hydropower plants connected to large reservoirs. This type of generation is flexible and very quick to respond to fluctuations in demand compared to other major generation technologies. All power generation systems are not 100 per cent effective and water levels in reservoirs are subject to weather conditions, and therefore electricity production varies throughout seasons and indeed years. During 2015, Ed Davey, the Energy Secretary, said the deal would give Britain access to Norwegian green hydropower at the flick of a switch, to replace wind turbines in the UK when the wind was not blowing, and would benefit both Britain and Norway. The proposed 350-mile cable was then estimated at a cost of £1.75 billion. It is claimed North Sea Link will enable both countries to maximise the use of these natural resources for the benefit of consumers in Norway and the UK. Such that when wind generation is high and electricity demand is low in the UK, the North Sea Link will allow up to 1,400 MW of power to flow from the UK, conserving water in Norway's reservoirs. But I would suggest this does not stand up to closer scrutiny as a number of variables have to link together for this to happen such as, during high winds over the UK and demand has to be low, whilst Norwegian demand has to be high. Murphy's Law will dictate that everyone will be in bed under such windy conditions
– then what happens when Norwegian reservoirs are full and UK power is not needed? It is also claimed that when demand is high in the UK and there is low wind generation, up to 1,400 MW (1.4 GW) can flow from Norway, helping to ensure secure electricity supplies. This is truly testing credibility as the UK demands an annual peak of 47.275 GW and an average of 29.653 GW – so even when providing its maximum power at
1.4 GW the interconnector will not make any significant impact especially when the UK is at peak demand. Readers should also note that at the start of 2022, the UK had a total installed wind farm capacity of over 24 GW, thus under a high pressure system, when there is little wind the 1.4 GW

interconnector will make little impression – it would appear that UK Government Energy Secretaries need an in-depth lesson in conventional and wind power engineering – complimented with a dose of common sense. The North Sea Link is not the only interconnector connecting Norway to another country as there is the undersea NorNed link between Feda in Norway and Eemshaven, Netherlands: it spans 580 kilometres (360 miles) delivering 700 MW of high voltage direct current power. It was once the longest submarine power cable in the world.

The Nordlink

There is also the NordLink, which opened in May, 2021 and provides a direct link between the Norwegian and German energy markets. It is a subsea 1,400 Megawatt (MW) HVDC power cable between Norway and Germany. It is over 500 kilometres (310 miles) long and operates at a voltage of 500 kV. It is said the high-voltage DC link will enable the exchange of 1,400 megawatts of renewable energy – wind power from Germany and hydropower from Norway. It would appear that Norway is planning to be the electricity hub of a number of countries, and at *face value* this would seem reasonable. But this system has an *Achilles Heel* in that if a number of countries demanded electricity simultaneously, for one reason or another, then one or more will have to suffer. Thus dependent countries will lose control of their energy requirements – not a comfortable position to be in. This is not fantasy and to reiterate yet again, the fishing row between Jersey and France clearly demonstrated friction during May, 2021, when the French, a NATO ally, threatened to cut off power to Jersey.

Greenlink

The Irish-Welsh Greenlink project is a proposed £453 million seabed and underground electricity interconnector planned for commissioning in 2023. The capacity of the link will be 500 MW and the project will connect EirGrid's 220 kV transmission sub-station at the Great Island, Wexford County, Ireland with National Grid's 400 kV transmission sub- station at Pembroke, Wales, UK, covering a total length of approximately 200 kilometres. Both the sub-stations will be interconnected by two 320 kV HVDC cables. The cables will make landfall at Baginbun Beach in Ireland and Freshwater West in Wales, and it is claimed the cables would carry enough power for 380,000 homes. The project also involves two converter stations, one near the Great Island sub-station, and the other near the Pembroke sub-station for the conversion of electricity from AC to DC and vice versa. Approximately 160 kilometres of offshore HVDC cables will be buried in the seabed of the Irish Sea, while approximately 22

kilometres will be underground on land. The HVDC cables will be buried in a single trench to connect the converter stations to the sub-stations, and has been selected over an AC connection, since AC is technically difficult over such distances as mentioned earlier. It should be noted the Pembroke sub-station is near the 2000 MW CCGT Pembroke Power Station. It is claimed the project will bring significant benefits on both sides of the Irish Sea relating to employment, energy security and the integration of low carbon energy sources. For Ireland, it provides a natural link to EU and Nordic electricity markets via Great Britain. Greenlink say that effectively plugging one country into another could help with future energy security. But I would respectfully suggest that would only be valid if countries, as we have mentioned earlier, do not fall out. The Greenlink Interconnector is designed to share energy between the National Grid and its Irish equivalent. But planners must have had their tongue in their cheeks saying it could help to reduce electricity costs on both sides of the sea and interconnectors have a proven ability to lower prices for consumers at both ends – REALLY - I must look at my power bill more closely in future! It is claimed there will certainly be a benefit from lower price Irish power, since Ireland has a significant number of renewable energy sources, so when the wind blows or when it is sunny they can export energy at a lower price into the United Kingdom, which will lower prices. To be sure, that sounds a lot of blarney to me and is something I will believe when I see it
– my power bill continues to rise – how is yours faring dear reader?

The North African interconnector

There are plans by the UK-based Xlinks to lay a 3,800 km (2,361 mile) HVDC submarine link between Guelmim Oued Noun in Morocco and Alverdiscott (between Barnstaple and Bideford), North Devon, to connect a 10.5 GW wind-solar complex in Morocco to the UK grid. The renewable energy complex will be linked to 5 GW/20 GWh of battery storage in Morocco. It is expected that 3.6 GW of electricity will be exported to the UK for at least twenty hours a day. The developers said the link would comprise of four separate cables and will be the longest seabed interconnector in the world. The cable will run along the Continental Shelf skirting Spain, Portugal and France. When fully completed it is claimed the project will deliver 26 TWh power to the UK each year – during 2019 the UK used 346 TWh of power. The link will provide power to the UK for a CfD price of around £0.048 per kilowatt-hour (kWh). This will be slightly cheaper than off-shore wind tenders which, at the time of writing, are about £0.040/kWh, although much less than the £0.0925/kWh for nuclear power from the Hinkley Point C nuclear power plant currently under construction. Note: Contract for Difference (CfD) is a form of financial investment that allows trading on a range of global markets - it is a contract between the buyer and seller that stipulates the buyer must pay the seller the difference between the current value of an asset and its value at contract time - it allows traders and investors an opportunity to profit from price movement without owning the underlying assets. It is claimed that cable losses along the entire transmission line are estimated between 10 per cent and 12 per cent, and it is said these losses are justified, by a very low LCOE for the solar and wind power plants in Morocco. Note: The Levelised Cost of Energy, or Levelised Cost of Electricity (LCO), is a measure of the average net present cost of electricity generation for a generating plant over its lifetime. The North African Interconnector is expected to have the first cable working by 2027 and a second cable two years later. The cost of the project will be £16 billion. As a comparison it should be noted the 2,100 MW, CCGT Pembrokeshire gas-fired, power plant, which opened in 2012, was built with an estimated investment of £1 billion and has a thermal efficiency of 60 per cent. It is one of the largest and the most efficient CCGT power plants in the UK. It is the author's opinion that for a safe, sustainable and viable economy the UK should not put itself in the position to be dependent on foreign countries for its electrical energy – it is vital we have security of supply. The reader is again reminded that during May, 2021 the EDF, the French-owned electricity company EDF threatened to turn off the power to Jersey. This 'Writing on the wall' should be a timely reminder for Government not to be dependent on other countries for UK electricity. To be sure, we are virtually sitting on a gold mine as beneath the country's surface are layers of shale, not to mention we are still surrounded by unexploited oil and gas reserves under the sea. With modern drilling techniques which offer environmental sensitivity, these untapped resources should be fully exploited by employing fracking. At this moment in time, UK electricity generation should be provided by efficient natural gas power stations supported by tidal energy and hydro schemes, until greener and sustainable energy sources can be realised as explained in this chapter.

Tidal energy

A powerful renewable energy source that has great potential around the UK is TIDAL ENERGY. This is a form of hydropower that converts the energy obtained from tides into useful forms of power, mainly electricity. Although not yet widely used, tidal energy has great potential for future electricity generation. It is deplorable that the billions being spent on large scale wind and solar have not been spent on the development of tidal power around the coast of Britain – a much better return on the money spent. Tides are more predictable and consistent than the wind or sun – it is sobering that wave and tidal stream energy has the potential to meet up to 20 per cent of the UK's current electricity demand, representing a 30-to-

50 GW installed capacity. For more detail readers may wish to go to:

https://www.gov.uk/guidance/wave-and-tidal-energy-part-of-the-uks-energy-mix

It is unbelievable that we are not exploiting the tides when it is estimated the UK has around 50 per cent of Europe's tidal energy resource. Studies have estimated the UK's tidal generation could produce between 12 per cent and 20 per cent of UK electricity. It should also be recognised and it is *important* to know that a tidal barrage or lagoon can be much more than a means of generating electrical power as we shall see later in this chapter.

Tidal schemes can fall into one of four categories:

1. The TIDAL BARRAGE, which is simply a dam for trapping water at high tide then releasing it through turbines to generate electricity.

2. A TIDAL LAGOON, very similar to the tidal barrage except a dam is replaced by a 360 degree enclosure that traps the water.

3. TIDAL STREAM GENERATOR, works very much like a wind generator by using water rather than wind, whereby sea currents turn the blades of the generator.

4. DYNAMIC TIDAL POWER, being a technology that uses the difference between the potential energy and kinetic energy of tides. It is a strategy whereby long walls are built out perpendicular to the shore line, and attached to these walls are walls running parallel to the coast line. During the movement of tides, water on one side of the wall perpendicular to the coast is at a higher level than the other side. As this water flows through the wall it drives a series of turbines installed within the wall to generate electricity. In the areas where these systems might be implemented is usually where tides flow parallel to their respective coasts. It should be noted that the walls are designed with bi- directional turbines, which flip 180^0 degrees after each tide in order to generate power both when the tide comes in and goes out. The added output from having bi-directional turbines is a huge advantage for these types of systems since they allow the power output to basically double.

The United Kingdom, from a tidal energy perspective, is extremely blessed in having the second highest tidal range on the planet in the Bristol Channel, and which begs the question as to why the UK Government rejected the Swansea Bay Tidal Lagoon in 2018? It is an ill-founded and very myopic decision - the prospect of building the world's first tidal power station in Swansea Bay would be momentous, offering the UK the chance to become a world leader in tidal power.

The concept of tidal lagoons and barrages are not just about generating electricity, as they can offer many other facilities and benefits, road and rail links, providing much needed employment and attracting tourism to the local area.

Swansea Bay Tidal Lagoon

During 2018 the Business and Energy Secretary Greg Clark said the £1.3 billion project was not value for money, despite claims by the developers Tidal Lagoon Power (TLP) that the UK has one of the best tidal range resources in the world. The Tidal Lagoon Power team had spent a long time looking at potential commercially viable sites of which six have been taken forward into their development programme. The tidal lagoon will generate 532 GWh per annum. What a supreme opportunity for Wales and the UK to lead the World in tidal power - this very short-sighted decision needs serious re-appraisal and should be reversed. Of course, the whole of the Bristol Channel should be considered for tidal energy and not just Swansea. The £1.75 billion North Sea Link Interconnector money, for example, would have been more wisely spent on the Swansea project, generating our own secure green energy, saving £0.45 billion, with the prospect of being a world leader in this technology. Another option would be to take £1.3 billion from the £14 billion Overseas Aid Budget, which gives money to rich countries that have *space programmes* and thus it follows, do not need such aid. Tidal lagoons can provide predictable, zero carbon electricity, with their construction taking years, not decades. Their output can be controlled in order to provide balancing services in conjunction with other sources of generation; a number of tidal lagoons in multiple locations around the UK coast, with different tidal cycles, can provide around the clock grid management services. As mentioned earlier, the billions already *wasted* on large scale wind and solar farms, in and around the UK, would have been far better spent on tidal generation. Although maximum tidal generation is not *constantly* available in *one* location, Britain has a very long coast, and tides are predictable, thus proving manageable for the National Grid, especially when compared to variable and unpredictable large scale wind and solar. The Swansea Tidal lagoon, featuring a U-shaped sea wall with turbines, operating 14 hours per day with a maximum output of 320 MW would be able to generate clean energy from the tides, capable of powering about 155,000 homes - about 90 per cent of homes in Swansea Bay. Why is Westminster not supporting this green energy project when 86 per cent of local people registered their support, with a 1000 residents signing up to join one of four independent Active Supporters Groups. Additionally, around 120 people from South Wales took up the opportunity to invest in the development of the tidal lagoon. It would appear the only significant opposition came from the Business and Energy Secretary – surprise, surprise! Technically challenged and myopic politicians need to recognise the Swansea project will be much more than a tidal power station. It will be highly beneficial to the Swansea Bay waterfront regeneration ambitions, affording recreation and amenity facilities. The seawall will be free to access for exercise or just a stroll to take in the spectacular views of Swansea Bay. There will be a playground, beach and rock pools, as well as art installations from local and international artists and the Offshore Visitor Centre, where visitors will learn more about tidal power and enjoy the truly unique seascape setting. The lagoon will be a fascinating place to visit, encouraging healthy living and engagement with the great outdoors. This is all the more true when you consider the wider opportunities created by the sheltered waters and features of a tidal lagoon. It will offer the opportunity for a whole range of local, national and international sports, including cycling, walking and running around the lagoon wall, not to mention sea angling, open water swimming, canoeing, rowing and sailing within the lagoon itself. There would be a fabulous boating centre (with access for disabled users) and permanent access to the water. Tidal Lagoon Power has liaised with many major sporting organisations which have expressed interest in using the lagoon's unique facilities to host events, including British Triathlon, British Rowing and the Royal Yachting Association. Swansea's schools and universities will also use the facilities for training and events. Regarding conservation and biodiversity the lagoon creates the perfect space – it has the potential to provide a lobster and oyster hatchery, and develop a marine aquaculture zone in partnership with local businesses to try new ideas and to help grow their businesses. There is consideration to include edible seaweed, cockle and mussel farming and further development work around integrated multi-trophic aquaculture (IMTA) – a way of a number of different species living together for environmental and economic benefit. The lagoon offers the prospect of restoring the native Swansea oyster population, a former Swansea industry that provided 500 local jobs and landed 16 million oysters per year at its peak in the late 1800s. The creation of a new seawall will offer opportunities to create new habitats to be colonised by a large variety of marine species, further encouraged by collaborative work on 'bioblocks' that can be placed along the lagoon wall. Enhancement work would also be undertaken to provide kittiwake roosts, new salt-marsh, dunes and grassland. The lagoon would support the development of world class research facilities, to inspire students and develop skills and knowledge around themes of tidal energy, coastal engineering and marine conservation.

The developers say it will boost the Welsh economy by an estimated £76 million a year, creating nearly 2,000 jobs during construction and 180 during operation.

Then, of course, there is Tourism. The tidal lagoon has the potential to become a major tourist attraction, attracting thousands of visitors per year, and much needed income for this part of Wales. The Welsh Government's tourism organisation 'Visit Wales' and local tourist and business groups see the lagoon as a magnet for visitors; in much the same way as the Eden Project has for Cornwall. This is not to overlook the hollowing out of a mountain in north Wales for the construction of the pumped-storage hydroelectric scheme near Dinorwig, Llanberis, Gwynedd, which is now known as the 'Electric Mountain' and attracts around 250,000 visitors per year. Looking further afield, a tidal barrage across the mouth of the Solway has the potential to produce 5.5 GW to 8 GW of power, whereby a barrage across Morecambe Bay could yield 3 GW, bearing in mind that is 1 GW more than the 2000 MW Pembrokeshire CCGT Power Station. The Bristol Channel Barrage has the potential to generate the same amount of electricity as three of the latest nuclear power stations. In other words the barrage could supply 6 per cent of the current electricity usage for England and Wales.

Northern tidal power gateways

There are plans to create a 14 km gateway and highway across Morecambe Bay between Heysham and Furness, and a 5.5 km crossing for the Duddon Estuary to improve road access to Cumbria's West Coast. More than 130 tidal turbines are planned to generate predictable renewable energy sufficient to power up to 2 million homes, establishing a new industry in the North West creating thousands of new jobs. The estimated   construction cost is £8 billion and will create over 7,500 construction jobs. Apart from generating between 7 TWh to 8 TWh (7 million to 8 million MWh) per annum it will reduce travelling distance by 50 per cent with a saving in travelling time by 75 per cent coupled with a consequent saving in fuel. It is estimated there will be 4.5 million annual crossings. The barrage roads will create a new transport link from the M6 in the south to the A595 north of Millom and will connect to each other via Furness. The Northern Tidal Power Gateways will incorporate up to 132 x 30 MW tidal range turbines, a total of 3.960 GW (3,960 MW), in prefabricated turbine housings to harness the vast renewable predictable energy of the large tidal range – up to 10 metres – in Morecambe Bay and the Duddon Estuary.   The tidal barrage across Morecambe Bay would not just generate electricity but, as mentioned above, incorporate a road bridge link built between Morecambe Bay and the south Cumbrian coast. The turbines would be designed to be inside the bridge, allowing for the production of emission free electricity. The expected service life of the tidal power barrage is up to 120 years. It will be capable of improvement and update over that period, and does not carry with it the horrendous and hazardous decommissioning costs a nuclear power plant of comparable generating capacity. The generation from the barrages is comparable with a Nuclear Power Station, and equates to around 7 per cent of the North West's electricity requirement, or 2 per cent of the country's requirement. As a direct comparison and to put things into perspective, Hinkley Point C Nuclear Power Station is a project to construct a 3,200 MW plant in Somerset, England. The power station at the time of writing has an estimated cost of more than £20 billion, so for the cost of Hinkley Point, nearly three similar Northern Tidal Power Gateways could be built, and capable of supplying GREEN energy - there would be no horrendous problems with sighting, decommissioning or the problems with *hazardous nuclear waste*. Personally I would not wish to live near a nuclear fission power station, and after a holiday in Cumbria and having spoken to many folk living in and around Whitehaven, this has simply confirmed and reinforced my sentiment.

A similar but smaller barrage with turbines across the La Rance estuary in Brittany has been running successfully since 1966, and a recent inspection has found little damage and relatively easy improvement prospects, particularly by using modern computer technology to control and maximise output of the turbines. At the time of writing it produced electricity for as little as £10/MWh. The 24 turbines reach peak output at 240 megawatts (MW) and average 57 MW, a load factor approximately 24 per cent, at an annual output of approximately 500 GWh. It is recognised that tidal barrages can have an environmental impact, but then all types of generation plant make an impact, with nuclear and coal being the worst. The 'trick' is to keep the impact on the environment to a minimum, and as the old saying goes, 'You cannot make an omelette without cracking eggs'.

Dragons Gate

Why is Government so myopic, bereft of any imagination - is it because they suffer zero technological, engineering and business skill and common sense, when a Bristol Channel Barrage offers so much more than just the generation of electricity - the scheme has the potential to offer a new all-weather road and railway crossing, and become a world tourist attraction, thus subsequently having the potential to pay for itself many times over, when all the factors are considered – it just takes the imagination and will. The project would have the advantage of the closing of the existing Severn

Railway Tunnel (Welsh: Twnnel Hafren). The tunnel was built by the Great Western Railway (GWR) between 1873 and 1886. Unfortunately during the building stages of the tunnel workmen hit what is known as the 'Great Spring' which necessitated the building of a huge pumping station at Sudbrook, Caldicot, on the Welsh side of the estuary. On average approximately 50,000,000 Litres (11,000,000 imperial gallons) per day of fresh (spring) water are typically being pumped from the tunnel; this is normally released directly into the adjacent River Severn. The huge pumping station once housed six 70 inch Cornish pumping engines to keep the tunnel clear of water. The steam engines of the pumping station were scrapped in 1968 to be replaced by electric motors – but should the reader be interested, the beam of one of the engines is located outside the Swansea Museum, Swansea. Additionally to the saving of the railway tunnel maintenance and cost of pumping, a new road crossing also could signal the closure and dismantling of the first Severn Road Bridge which was opened on 8 September 1966. To further enhance the scheme half way across the barrage an almond shaped island could be constructed and incorporated into the barrage, with the narrow ends of the almond island pointing up and down stream. The island would become a large visitor centre with a railway station and hotels next to a large car park offering motorway services coming off the road crossing. A custom built building could offer a 'Museum of the Severn Estuary' coupled with access to view and offer guided tours of the turbines comprising the generating plant, in much the same way as the Electric Mountain in Dinorwig, North Wales. The island should be of sufficient size to accommodate other tourist attractions such as shops, cafes and a small harbour to offer boats trips up and down the Severn Estuary. It is recognised that most visitors would obviously be attracted during the summer months, but with the road crossing being integrated into the motor way system and a railway station, the potential for an all year round attraction would be easily accommodated. As an example, a special Christmas Severn Estuary Festival could be held on the island, and I am sure that many entrepreneurs will have lots of other ideas. The entrance to the Barrage on the Welsh and English side of the Bristol Channel would lend themselves to the construction of impressive dragons on top of large arches. On the English side would be a large Red Dragon, atop its arch, welcoming travellers over the barrage into Wales (Cymru). On the Welsh side would be another large arch with a Green Dragon (St George's Dragon), welcoming travellers into England. Both Dragons would be lit at night and the crossing appropriately named DRAGONS GATE. To be sure, the dragons on top of their arches have the potential to become a tourist attraction by themselves - much the same manner as the Angel of the North. The planning and construction of such a project would create much needed jobs on both sides of the channel, and once completed would continue to create numerous jobs as a visitor attraction. The whole project could become a wonder of the 21st Century, and folk would visit from across the globe – it simply takes the political imagination and will to get the job done. The volume of water that passes over the Aberthaw-Minehead barrage line in the Bristol Channel, as it flows back to the sea, is much greater than that passing over the Lavernock Point to Brean Down barrage line. The area covered upstream of the Aberthaw barrage is more than 14.5 to 11.0 times greater than that upstream of a Lavernock barrage. Combining the tidal height during electricity generation and the associated flow-rate shows a barrage between Aberthaw and Minehead has the maximum potential for generating electricity in the Bristol Channel/ Severn Estuary. A further significant benefit would be the protection of flooding for all the Severn Estuary above the barrage – such as Cardiff, Newport, Gloucester, Avonmouth, Bristol, the Somerset Levels, and most importantly the Hinkley Point C Nuclear Power Station. It should be clearly understood that the Global potential for tidal power is massive with the World Energy Council estimating that up to 1,000 GW of marine energy could be installed by mid-century – equivalent to half of the world's present coal capacity. For all those myopic critics of tidal energy and bearing in mind waste of money, it is *instructive* to recollect the mind boggling series of botched Government IT projects that cost taxpayers more than £26 billion for computer systems that have suffered severe delays, run millions of pounds over budget or have been cancelled altogether. For example, the Government handling and the abandoned NHS IT patient record system in 2002 cost the taxpayer nearly £10 billion, for what would have been the world's largest civilian computer system. According a highly critical report from parliament's public spending watchdog, the cost was likely to be several hundreds of millions of pounds higher. Then of course there was the scandal of the Post Office Horizon computer system and its widespread miscarriage of justice. This resulted in the Post Office Chief Executive announcing in April, 2021 that the £1 billion Horizon system would be replaced with a new cloud-based IT system. It is scandalous when you consider that the wasted billions could have been spent on tidal schemes such as the Northern tidal gateways - creating thousands of jobs during construction, reducing land transport mileage and repaying the capital many times over.

Floating platform marine current power

An interesting commercial tidal project is being carried out in Orkney employing a tidal-powered electricity generator, which the makers, Orbital Marine Power, Kirkwall, Orkney, say is the most powerful in the world, has started to generate electricity in Orkney during July, 2021. The innovative, floating 680-tonne generator, called 'The Orbital 02' is anchored in the Fall of Warness where a subsea cable connects two 1 MW offshore generators to the local onshore electricity network. It is claimed the generation capacity of 2 MW will meet the annual electricity demand of around 2,000 homes for the next 15 years. Orbital's unique floating platform is moored via anchors in powerful tidal streams, but of course will lend itself to any situation where there is fast flowing water such as river currents. Underwater rotors capture the dense flowing energy, and I would agree with the makers that this enterprise marks another major step forward for the UKs nascent marine energy sector.

Floating Offshore Wind (FLOW)

The reader will not be surprised that I have little confidence in the concept of floating wind generators far out at sea. Every sailor knows that violent storms at sea take no prisoners, and the Atlantic and North Sea coasts are testimony to the numerous shipwrecks over the centuries by an unforgiving sea. So it was with dismay that I read the UK government has made a commitment to reach 40 GW of offshore wind capacity by 2030. To the extent it is claimed that floating offshore wind can be deployed in deeper waters, where wind resource is higher and is not constrained by the need for fixed bottom foundations. I would agree that the wind resource is higher out at sea, and as such I wonder how often these floating marine wind generators will have to shut down because of very high winds offering the prospect of little or no generation – never mind the consequences of violent storms out at sea – let us hope the platforms, which are anchored to the seabed by means of flexible anchors, chains or steel cables are up to the job. Blue Energy, a pioneering Celtic Sea energy developer, announced in March 2021 it is to develop a 96 MW Erebus floating wind project, in the Welsh waters of the Celtic Sea. Hopefully the name of the project will not prove prophetic and unfortunate - in ancient Greek mythology the word means a place of darkness between the Earth and Hades. More unsettling is that HMS Erebus was one of the two ships and crew that disappeared with the Franklin expedition, while searching for the Northwest Passage in 1848.

Natural gas power stations

The Government is blindly committing the UK to a large dependency on solar and wind generation - each are prone to unpredictable reductions in output, such as when the wind suddenly drops or it is cloudy, which makes for difficult Grid control. The situation is very challenging during a cold and frosty winter night when there is no wind or solar radiation. Where does the electricity come from under these conditions to ensure the lights stay on? Of course, it comes from the National Grid which is predominantly dependent on fossil-fuel generation in the form of gas. But it is nothing short of scandalous that the UK has very limited capacity to store gas when compared to countries such as The Netherlands which has nine times the storage capacity, and Germany has sixteen times. It was reported in the media during January, 2022 that the energy giants had doubled the amount of gas extracted from the North Sea and Irish Sea to foreign buyers when UK families are struggling with higher bills – this begs the question why? Currently Britain produces about half the gas it uses domestically from fields under the North Sea and Irish Sea – but could be totally independent if it started fracking for gas. With natural gas there is less pollution than oil or coal producing 60 per cent less carbon dioxide, thus producing more energy per unit CO_2 emitted. The world has plenty of relatively cheap natural gas, and the UK has the potential to produce its own gas with hydraulic fracturing, informally referred to as 'fracking'. This is simply a 'well development process' that typically involves injecting water, sand, and chemicals under high pressure into a bedrock formation via a well, see later in this chapter. It should be noted that producing power via gas is cost-competitive with coal in the U.S., and recently has been replacing coal to produce electricity. This has resulted in the U.S. decreasing its annual CO_2 emissions by nearly 10 per cent.

The fossil-fuel natural gas power station is the far cleaner and efficient option than oil, or coal power plants. Until greener power stations such as Fusion plants can be commercially realised, then power stations such as the Combined Cycle Gas Turbine (CCGT) power station of 1,500 MW capacity or greater, should be built and then eventually phased out with the introduction of genuine greener production methods. A combined-cycle power plant uses both a gas and a steam turbine together to produce up to 50 per cent more electricity from the same fuel than a traditional simple-cycle plant. The waste heat from the gas turbine is routed to the nearby steam turbine, which generates extra power, thus improving performance. CCGT power stations have the ability to improve the environmental performance of fossil-fired power plants - their flexible operation and reduced environmental impact, improve energy efficiency and reduce atmospheric emissions. The Pembrokeshire 2000 MW Power Station operates a CCGT process to produce electricity. At the power station, air goes through a compressor and is mixed with natural gas in the combustion chamber and burned. It is important to note that in addition to reducing the amount of fuel required, CCGT power plants emit only half the amount of carbon dioxide and one-third the amount of nitrogen oxides, and virtually eliminates emissions of sulphur dioxide compared with conventional fossil-fired generation resources such as coal-fired plants.

Currently the UK dominant fuel consumption for the production of electricity is natural gas. As an example, on Sunday 30th January, 2022, at 15.55 GMT, a windy day, the total UK demand for electricity was 35.154 GW, of which the lion's share was gas at 11.037 GW; nuclear contributed 5.186 GW and wind 9.984 GW, whilst unsurprisingly solar contributed 1.26 GW, Coal at 1.121 GW, with the balance from other sources - with acknowledgements to www.gridwatch.co.uk.

Current energy provision

The closure of coal-fired stations with many nuclear power stations reaching the end of their life coupled with the growth of wind and solar farms will certainly put UK power supply in jeopardy. Could this be one of the reasons Government are pushing for every home to have a Smart Meter fitted so as to quickly and remotely isolate customers when the going gets tough? Limited wind farms (too little or too high a wind - no power) and restricted solar parks (no solar radiation - no power) are not going to save the situation, far from it. If we covered every acre of ground in the UK with thousands of wind farms and solar parks, then on a cold and cloudy mid-winters day, with little wind blowing, where on earth will the power be coming from – perhaps 'Alice needs to ask the Mad Hatter'
– it is as daft as that!

The country desperately needs a Government that has the technical and engineering knowledge to call an immediate halt to nuclear fission power stations, large scale wind farms and solar parks. The money consequently saved being channelled to natural gas power stations, tidal energy and hydro schemes in the UK. It is an utter disgrace that billions have been squandered on the unpredictable and limited generation that wind farms and solar parks can offer. It is very sad that the £1.3 billion Swansea Bay Tidal Lagoon has been scrapped by the Business and Energy Secretary Greg Clark. It was reported that the decision was (rightfully) slammed by local politicians across the parties - let us hope wiser heads will prevail and the scheme will eventually obtain approval.

Pembrokeshire CCGT Power Station in West Wales began commercial operation in September 2012; it was built with an estimated investment of £1 billion which is obviously far much cheaper than the 20 Billion estimated to build the new proposed nuclear Hinkley Point Power Station in Somerset. The Pembrokeshire Power Station has a total capacity of 2,100 MW, and is enough to power around 4 million homes – more than twice the number of households in Wales. At the time of writing, the state- of-the-art CCGT Pembrokeshire Power Station is one of the largest and most efficient plants of its kind in Europe - it is also the largest power station to be built in the UK since Drax Power Station came online in 1986. It produces less than half the CO_2 emissions compared to a coal- fired power station, and its flexible technology means Pembrokeshire Power Station is able to respond quickly to the market to provide highly flexible and reliable power to meet the UK's demands. It cannot be stressed enough times that it would make sense for UK shale gas extraction to start immediately. The world has plenty of natural gas and if the UK had a viable natural gas industry then it could 'play' the world market such that, if and when, import gas is cheaper, it would make sense to then buy foreign gas and preserve our own, having the best of both worlds as they say.

It is worth echoing that a return to dirty coal-fired power stations would be a retrograde and unacceptable step. It has been demonstrated that large scale wind, solar or tidal generation, individually or combined, will not satisfy UK peak demand, or total annual consumption, due to their limitations. Large scale wind and solar generation is a waste of time, space and money, and nuclear fission power stations are horrendously costly, not to mention their decommissioning and nuclear waste insurmountable problems – there is also the possibility/probability of a catastrophic meltdown - Government should learn the lesson from the Russians, Japanese and others! Therefore UK power strategy should, unquestionably, be based on relatively low-polluting, efficient, cheap and quick to build gas-fired power stations of the CCGT type - the Pembrokeshire 2100 MW Power Station is a typical example. This 'core' generation should be complemented with reliable tidal energy around the UK coast, recognising that the tides occur at different times along the coast. This generation mix should also include hydro-electric plants and pumped-storage such as Dinorwig in Wales. Unless we plan to sell power to other countries, at a profit of course, there will be no need to build costly Interconnectors as we will be self-sufficient in power needs.

We are where we are, and our technological dependent society will collapse without a sustainable and secure power supply.

That is a certainty and we ignore this fact at our peril. Our technological society depends on a reliable source of power. If the Grid fails, and without local battery or generator backup, then so do the banks, vehicle service stations, supermarkets, hospitals, television, radio and entertainment venues, all central heating systems pumps and control devices. Mobile telephones need their batteries recharging, microwaves, electric cookers and ovens, street lighting, air terminals will not function without electricity and aircraft will be grounded, electric vehicles will prove useless. The list is endless and our society will quickly grind to a halt and collapse - that is the brutal reality! We should all be experiencing restless nights at the inept and cavalier way UK Government is handling this vital energy requirement. It will be too late when the lights start going out, as we will then have to face a nightmare scenario - it is imperative dear reader that you do not underestimate the gravity of the situation. But you are not helpless as you can make the difference, as there are local councillors and MPs to make representation to. Indifference, apathy, lack of interest and unresponsiveness will bring its own rewards.

Hydrofracking (Shale gas)

There has been a lot of misleading information written about the subject of fracking ranging from the poisoning of drinking water, to the collapse of buildings as a result of *fracking* induced earthquakes. It is amazing how many people will publicly criticise an issue when having little or no understanding of the issue they are protesting about – it would seem that unwarranted FEAR rules the day in the 21st Century. Take the case of contaminating groundwater with methane, fracking fluid, chemicals, and dissolved contaminants in flow water due to the activities of shale gas operations – do critics not realise that ground water is near the surface whilst shale gas lies some 2 kilometres to 5 kilometres below the surface – so unless the integrity of the drilling pipe (casing) fails, there is little chance of ground water becoming contaminated. To put this into some kind of perspective, how many know the odds are greater for a person in London to be blown up as a result of a leaking gas pipe whilst walking along a pavement, than by accidently drinking water that has been contaminated by fracking. Services such as telephone and electricity cables, water and gas pipes are buried under the pavement at depths better measured in inches than feet, and a lot of these services are now many years old and are failing.

The Geological Society of London will inform you that shale gas is extracted from fine-grained sedimentary rocks (called shale) formed over millions of years as the result of compaction of fine particles of mud – organic matter can be trapped in the layers as they are compacted, and are slowly converted through heat and pressure into hydrocarbons such as natural gas. The main component of shale gas, and the primary reason for its extraction, is methane, which makes up between 70 per cent and 90 per cent of shale gas together with smaller amounts of other light hydrocarbons, carbon dioxide, oxygen, nitrogen, hydrogen sulphide, radon and rare gases. There are large sedimentary basins in the UK which contain significant shale sections. Exploration for shale gas in the UK is still at an early stage, so there is currently no clear consensus about how much shale gas is under the ground and the prospects for extracting it economically. Nevertheless, most geologists agree that there are reasonably significant onshore resources. But whether these resources are exploited will depend on economic, environmental, social, and regulatory constraints. The Geological Society of London point out that there are risks and challenges associated with the extraction of any mineral resources, including shale gas – thus it is important that such activity is appropriately regulated, and risks identified and managed. Three areas of potential risk which have given rise to particular concern among policy- makers and the public are: groundwater contamination; water sourcing and disposal; and induced seismicity. In the UK groundwater provides 35 per cent of drinking water, and is important in supporting surface water flow and regulating the health of ecosystems. Concerns have been raised about the possible contamination of groundwater by methane, fracking fluid chemicals, and dissolved contaminants in flow-back water, as a result of shale gas operations. If wells are properly constructed, contamination of ground water through migration of methane and fracking fluids from shale formations to shallow aquifers through stimulated fractures could only take place if the fractures are able to propagate vertically through the intervening layers of rock. Recent analysis of fracking operations in the USA, combined with data obtained from natural fracturing of rocks, indicates that the probability of a stimulated fracture exceeding a height of 350 meters is around 1 per cent. The analysis suggests that if a separation distance of at least 600 metres is maintained between aquifers and fracture zones, the risk of a fracture propagating to the aquifer and causing contamination is extremely low. There are recorded instances of methane in groundwater in the USA in areas where shale gas operations have taken place. But a more likely cause than migration through fractures is methane leakage at the well site itself, due to poor design or construction, or subsequent damage, (Historically, onshore U.S. hydrocarbon operations have not always been effectively regulated, and in some areas there is a lack of records relating to well design and construction). It should be noted that methane can also occur naturally in shallow groundwater. Geochemical analysis can distinguish this from thermogenic methane from deep shale formations. Induced seismicity is the release of energy stored in the Earth's crust triggered by human activity, and is known to be caused by activities such as mining, deep quarrying, geothermal energy production and underground fluid disposal.

Fracking
Hydraulic Fracturing

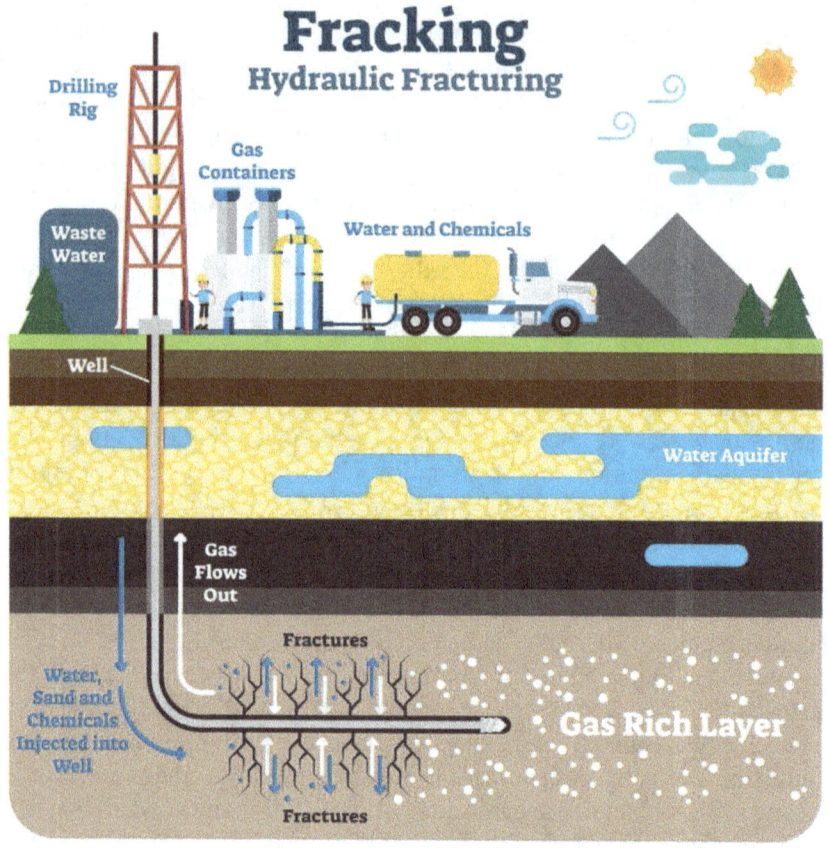

During 2011, two seismic events of magnitude 2.3 and 1.5 took place in Lancashire, close to a fracking test site operated by Cuadrilla. Operations were suspended, and subsequent studies have suggested that hydraulic fracturing is likely to have been the cause, by reactivating an existing fault. The crust in most of the UK is relatively weak, and unable to store sufficient energy for large seismic events.

This means that the largest natural earthquake we can expect is likely to be no greater than magnitude 6. However, based on our understanding of the mechanical strength of shale and case studies of fracking operations in the USA, it is extremely unlikely that seismic events induced by fracking will ever reach a magnitude greater than 3. These are likely to be detectable by few people and are highly unlikely to cause any structural damage at the surface. To minimise the risk of seismic events even at this level, operators should avoid drilling through or near faults, and microseismicity should be monitored in real time before, during and after fracking, with effective management systems in place to respond to the results, including monitoring possible damage to well integrity. The reader should recognise that there are in the region of 200-300 earthquakes in the UK every year, although the majority are so small that nobody is aware of them with only about 10 per cent detected by people each year. Nevertheless, a number of UK earthquakes have caused damage with the largest known earthquake (with a magnitude of 6.1) occurring near the Dogger Bank in 1931 – although it occurred 60 miles offshore it still caused minor damage to buildings on the East Coast of England. During 1884 the most damaging earthquake took place in Colchester where over 1000 buildings needed repairs with chimneys collapsing and walls cracking. Recently, during February 2015, an earthquake of magnitude 2.9 occurred in the Channel Islands which resulted in more than 100 reports from alarmed residents. The British Geological Survey said the epicentre was located about 16 miles south-west of St. Helier. Until we truly put an end to our dependence on fossil fuels and to resolve the current energy crisis facing the UK, fracking has to be more than an acceptable proposition. It has the potential to transform the economy of the nation, and at the same time, erode our dependence on imported natural gas. It is truly unbelievable that shale gas is taking so long to take off here in the UK, while in the USA they have grabbed this 'bonanza' with both hands and transformed the American economy. It is very revealing in 2019 when Natascha Engel, the then Government Commissioner for Shale Gas, resigned as a result of her dismay in Ministers jeopardising Britain's energy security, simply because they would rather appease noisy green campaigners than listen to scientific advice and that government policy is strangling the UK shale gas industry at birth – despite overwhelming *scientific evidence* that fracking, if properly regulated, is *totally safe*. Although I am not a supporter of any individual political party, I do salute Natascha Engel's courage, integrity and perception – noting that Natascha Engel was a Labour Party Member of Parliament (MP) for North East Derbyshire from 2005 until 2017 - and that it was a Labour Government that introduced the FIT payment at over fifty pence a unit for roof-fitted solar power generation - what is there not to like with such commitments, especially when the Conservatives have abolished the FIT payment, and have their heads in the sand over fracking.

Heat pumps

Fundamentally the heat pump is a system which transfers heat from outside to inside a building, or vice versa. For domestic heating purposes an Air-Source-Heat-Pump (ASHP) absorbs heat from external air and releases the heat energy inside the building, as either hot air, hot water- filled radiators, underfloor heating or domestic hot water supply; the same system can often do the reverse in summer, cooling the inside of the house. At first glance heat pumps appear a feasible and sensible answer to heating our homes, but unfortunately they have their short-comings, such as not being able to compete, in the majority of cases, with a conventional domestic central heating boiler.

Unlike the proposed electric vehicles (EVs) where it is difficult to estimate how many vehicles will be simultaneously charging at any one moment, and therefore how much electricity will be needed, it is much easier to ascertain for heat pumps. The calculation is made that much easier simply because during a cold, dark and windless winter's night every houshold will have a heat pump running (should they become a reality) to keep warm, that is, apart from empty dwellings - and possibly the odd igloo. During 2017 there were 27,227,700 households in the UK. Thus to realise a meaningful estimation, a calculation based on 20 million households using 1 kW of electricity revealed that an absolute minimum of 20 GW will be required. During 2020 the peak demand in the UK was 47.275 GW. Therefore to satisfy the requirement for 20 million households would mean realising a new peak demand of 67.275 GW of power, if both peaks coincide. This must be regarded as a conservative figure for it is realised that for many households 3 kW of heat delivered by a heat pump will not satisfy the heating and hot water needed. Additionally, it should be noted that as temperature decreases, so will the efficiency of heat pumps, which is unfortunate for cold winter use, as obviously that is when the heat is needed most. To this extra electricity demand needs to be added the significant additional electricity for the intended 40 million EVs that Boris Johnson envisages to obtain his zero carbon target – this target though is deemed a nonsense as concrete production alone makes up 8 per cent of carbon emissions – will we stop building and repairing altogether?

The heating performance and efficiency of an air source heat pump system is commonly measured by the Coefficient of Performance (CoP). The CoP is a simple calculation which works out how much energy the heat pump is able to extract from the energy source compared to the amount of electrical energy it uses. Generally speaking, the higher the CoP figure, the greater the efficiency of the heat pump. A CoP however only applies to a specific temperature, which means that the CoP rating is not representative of the performance that could be achieved across a whole year. A far more accurate assessment of efficiency therefore is provided by the Seasonal Coefficient of Performance (SCOP). It defines the performance of the heat pump over the course of the year, with seasonal variations in conditions.

TYPICAL AIR-TO-AIR HEAT PUMP

According to Worcester (Bosch Group) their Greensource air-to-air heat pump offers highly efficient heating in all seasons, with an industry leading SCOP of 3.8. Worcester also state there is a maximum heating output of up to 6 kW, which is sufficient for heating an area of up to 100 m^2 and it is this performance coupled with the highest levels of quality and reliability which are synonymous with the Bosch name throughout the world that persuaded me to purchase the Greensource air-to-air heat pump. Apart from allowing consumers to efficiently heat their homes, the Greensource air-to-air heat pump offers a cooling feature which ideally complements roof-fitted solar panel systems during a hot summer day when solar generation is at a maximum – in a sense it offers air conditioning for free.

The Worcester Greensource air-to-air heat pump is suitable for a wide variety of property types and sizes, and can *complement* existing gas, oil or renewable hot water systems. It can also offer stand-alone heating for the home in certain applications. A Greensource air-to-air heat pump was installed at our property at the end of January, 2014 and the total cost for purchasing and installation at the time was £1991.85 - it has certainly saved money in both electricity and oil (used for heating) and as such I would certainly recommend the fitting of one in a suitable situation – such as complementing roof-fitted solar panels. I should point out to the reader that the Greensource air-to-air heat pump did not have a weekly programmer built-in, although it did have a 24 hour programmer offering timer on, and timer off.

Readers who are not familiar with the technology of the heat pump can simply look upon the device as a 'black box' with its inputs and output. There are two inputs, namely:

1. Input for Grid or other sourced electricity.
2. Input for heat energy from the atmosphere, ground, pond, lake, or aquifer et cetera.

The black box has only one output and that is for heat energy in the form of warm air or water to heat a dwelling. As mentioned earlier, a heat pump is a device that 'transports' heat energy from one place to another place. This is the basic feature of how heat pumps work - an *air conditioner* is a form of heat pump in that it 'extracts' heat from indoors and pumps it to the outside. Thus on the indoor side you have cool air blowing out of the vent, after passing through a heat exchanger. On the outdoor side you have warm air blowing out of another heat exchanger. The heat exchanger on the indoor side is called an evaporator and the heat exchanger on the outdoor side is called a condenser.

Air-to-air heat pumps are suitable for a wide variety of property types and sizes, and as a generalisation can complement (not replace) existing gas, oil or renewable hot water systems. In a sense a heat pump when in heating mode can be seen simply as an ENHANCED ELECTRICAL HEATER with the operational electricity being supplied from the mains or private generation - with extracting heat from the outside air, ground or water source - which simply aids in the reduction of mains or private electricity.

Ground-Source-Heat-Pumps (GSHPs)

The same as Air-Sourced-Heat-Pumps (ASHPs), except the heat energy for these pumps is extracted from the ground. The ground sourced method of extracting heat is known as geothermal heating. The burying of the necessary pipe works will be a challenging, if not impossible task, for the majority of existing properties, especially high rise flats. Although the big advantage of obtaining heat from the ground is the temperature a few metres below the surface of the Earth keeps at a fairly constant level of about 10 degrees centigrade throughout the year. This heat, by means of suitable underground pipes and heat pumps can be extracted for space heating in buildings, and in some cases, pre-heating domestic hot water. By utilising ground-sourced heat the requirement, as in the case of air- sourced heat, reduces the use of mains electricity. A geothermal heating and cooling system is comprised of two distinct parts: a heat exchange unit that is installed in the interior of the home and an earth loop that is directly buried in the ground near the home, or in a body of water located by the home. The earth hose loop is made up of high-density polyethylene which is very durable and able to stand up to the elements it will encounter. This flexible hosing is installed either vertically, horizontally, or in a coiled fashion to absorb or disperse heat through the medium it circulates - geothermal heating is achieved when the system takes heat from the ground or water source and transfers it into the home through the heat exchanger - unit controls must be set to heat to achieve this. Likewise, when the switch is in the cooling mode, the system takes heat from the inside of the house and transfers it to the outside through the hose lines. As the fluid passes through the lines and the cooler ground or water, it returns to the heat exchanger to cool the home. The system uses a mixture of water and antifreeze (much like a motor vehicle radiator) to heat and cool the home.

Benefits of a geothermal system

- If the system loops are professionally installed, they are guaranteed for more than 50 years of trouble-free use.

- The units are very reliable and many have been in use for 30 years.

- The systems are safe to run because there are no open flames and therefore no danger of fires or carbon monoxide poisoning.

- Contributes to a healthy home, as those who suffer from dry, irritated nose or throat, will benefit as the units provide for a constant controlled humidity and temperature throughout the year.

- Simple operation allows switching from heating to cooling with a flip of a switch.

Therefore heat pumps, where they can be fitted, will complement central heating systems such as those run on solid fuel, gas or oil, and thus offer a saving on overall heating costs. Having fitted an air-to-air heat pump and solar panel system, I can speak from experience to the limited merits of the fitting of such devices to an existing house. During very sunny days during the summer months it was very welcome and satisfying to run the air-to- air heat pump in air conditioning mode, and knowing the electrical power was coming from the roof-fitted solar panels and not from the mains supply. Additionally, during the spring, winter and autumn months, on days when there was adequate solar radiation, it was pleasing to run the air-to-air pump as a heater with power coming from the roof-fitted solar panels. This did reflect in a saving of oil for the central heating system – there is no doubt that the use of solar panels and a heat pump to compliment a conventional heating system has its merits. But having said that, it should be noted that the solar panels or heat pump together, or individually, would fail to adequately heat the house, and certainly there would be no hot water without the assistance of the conventional central heating boiler. With regard to conventional central heating systems, that pumps hot water around house radiators and heat exchangers in hot water storage tanks. It is very important to realise they operate at relatively high temperatures. Thermostats control the output temperature for radiators and the hot water tank respectively. The recommendation is to set radiators at 75 degrees Centigrade (167° Fahrenheit) and the hot water tank at 60 degrees Centigrade (140⁰ Fahrenheit) to maximise efficiency.

On Monday 5ᵗʰ April, 2021, Boris Johnson, the British Prime Minister, and the Conservative Government set a goal of replacing 600,000 domestic boilers with heat pumps, every year, by 2028 to assist in decarbonising home heating and thus reduce the UKs greenhouse gas emissions. It is truly unbelievable that government departments dealing with energy do not realise that heat pumps, by their very nature, cannot directly replace conventional boilers for the same required/desirable comfortable level of heat in the home, simply because they function at lower temperatures and are dependent on two energy sources: mains electricity, and an air or ground thermal source. Note: where ground thermal source is constant the *variable* air is not. To try and obtain the same level of heating of a conventional boiler, such as a gas boiler, would necessitate in replacing existing radiators with larger ones, to spread the (lower) heat more quickly. Improving house insulation and possibly have access to another complimentary heat source. If a house that is dependent on an air heat pump, would wish to have an inside temperature of 24 degrees Centigrade, when the outside air temperature is say 4 degrees Centigrade, then obviously only 4 degrees of external heat is available. The balance of 20 degrees for heating will have to come from another source. The only other energy source available to the heat pump is that of electricity from the mains, or other generation source. Remember as the outdoor temperature drops, so does the heat output of the air source heat pump, which simply means the heat pump (the refrigerant unit) has to work harder.

I wonder how many Ministers realise that in places such as Colorado in the USA, where winter lows can drop to minus 7 degrees Celsius, homeowners often ask if a heat pump can really heat their home, when there is barely enough heat outdoors to extract. Heat pumps definitely lose capacity as the outdoor temperatures get colder, and in Colorado heat pumps have a *backup* plan for when that happens. When the outside temperatures get so low that the heat pump cannot extract enough heat to warm the home, the unit switches to *backup* heating – known as *supplemental* or *second-stage* heat. The heat pump system does this automatically - supplemental heat is provided by electric resistance coils, similar to those in an electrical toaster. In Colorado if the home has natural gas, the supplemental heat will probably come from a gas furnace that will assist when the heat pump is struggling. The gas furnace is far more efficient than the electric resistance coils - but whether the home uses electric resistance heating or gas, the supplemental heating is less efficient and therefore more expensive to run than heat pump heating. Another hurdle to the adoption of heat pump will be the cost of installation. At the time of writing I understand the cost for a heat pump that draws heat from the air would be in the region of £4,000-£9,000 and around £14,000-

£19,000 for one that extracts heat from the ground. Whether Government will offer *meaningful* subsidies remains to be seen.

The Government has foolishly set a target of reaching net-zero emissions by 2050, which means that the UK will be at net-zero when the amount of greenhouse gas emissions produced, is equal to the amount that we remove from the atmosphere. As the country's carbon emissions come from fossil-fuel boilers in homes, fossil-fuelled power stations, land, air and sea transportation, building industry et cetera, it is difficult to see how this can come about. The British Prime Minister, Boris Johnson, has acted in a rash way by announcing that gas or oil boilers will not be installed in new build homes from 2025, and there are also proposed plans to phase out the installation of new fossil-fuel boilers in general from the mid-2030s. What *means* will be heating our homes in the future as heat pumps or solar panels are not practical contenders - what other options are there?

Hydrogen domestic boilers

A plausible means to reduce natural gas used in domestic boilers is to mix it with hydrogen and this would reduce carbon emissions as hydrogen does not produce carbon emissions when burnt. Thus a hydrogen blend could be delivered to properties, incorporating a gradually increasing amount of hydrogen via the existing gas network. It is suggested that the level of hydrogen in the gas supply would be no more than 20 per cent providing 20:80 in favour of natural gas through the existing gas network. The adoption of such a scheme could prevent the need for millions of existing boilers to be removed. Therefore, over time, phasing hydrogen into the natural gas supply will have the effect of reducing the UK's carbon emissions coupled with little impact on homeowners. Indeed, a hydrogen boiler would operate in the same way as a traditional boiler, but would not produce any carbon emissions. Globally, there are a number of projects searching for an efficient way to produce low-carbon hydrogen. The UK government has invested £20 million in the Hydrogen Supply Programme to investigate how viable a hydrogen network would be. During 2019 a hydrogen gas trial (HyDeploy) was carried out at Keele University (near Newcastle-under-Lyme, Staffordshire), which introduced a 20 per cent hydrogen mix to the 130 homes and faculty buildings at the university, with no ill effects. Larger trials in the North of England are on track to go ahead in the early 2020s. A 20 per cent hydrogen mix across the entire network would work with current modern boilers, and would save six million tonnes of carbon emissions – equivalent to taking 2.5 million cars off the road. Hydrogen boilers would be installed in a similar way as a conventional gas boiler and would look very similar; they would be connected to the existing gas network. A lot of the internal parts would be identical, and only a small number of components, such as the flame detector and burner, would need to be replaced to be compatible with hydrogen. If the gas network should be converted to a 100 per cent supply of hydrogen then Gas Safe registered engineers would be trained to install them. Most modern boilers would be modified relatively simply to work with 100 per cent hydrogen. Therefore customers will not have any significant expense or inconvenience; in addition, installers will only need a small amount of additional training. As the hydrogen-ready boiler can heat a home on natural gas, a future switch to hydrogen will enable the boiler to operate on 100 per cent hydrogen after a short visit from a heating engineer. While current gas boilers will not work on 100 per cent hydrogen, most will work on a hydrogen blend. Worcester Bosch unveiled their very first hydrogen boiler in November 2019. As one of the leading UK manufacturers of gas and oil boilers, Worcester Bosch are now working to find low carbon alternatives and have stated that all boilers installed from 2025 should be hydrogen-ready. Boiler manufacturers such as Baxi and Worcester Bosch have already developed hydrogen-ready boilers and are calling on the government to mandate that all new boilers installed from 2025 be hydrogen-ready. The reader should be aware that hydrogen production is expensive as producing hydrogen in large enough quantities to meet demand is not cheap - hydrogen can be produced either by electrolysis, i.e. using electricity to split water into oxygen and hydrogen, or by Steam Methane Reforming (SMR). The latter method utilises a chemical process which is used in industry to produce hydrogen on a large scale and is fundamentally based on two chemical reactions which *ultimately* converts water and methane (natural gas) into pure hydrogen and carbon dioxide. It should be noted that natural gas reforming does not completely do away with carbon dioxide emissions, but simply reduces emissions when compared with burning natural gas in conventional boilers. Therefore in an effort to overcome this issue, low- carbon production methods need to be employed, and as an example, electrolysis would need to be powered by tidal and ocean current electricity generation.

Hydrogen fuel cell boilers

Basically a hydrogen fuel cell is an electrochemical device that converts the chemical energy from a fuel such as methane in natural gas into electricity through a chemical reaction with oxygen. The fuel cell has two electrodes, namely, an anode and a cathode. The cathode is negatively charged and the anode is positively charged. It can be said that fuel cells are similar to a battery except they do not run down or need recharging, simply because they produce electricity and heat as long as fuel is supplied. Water flows out of the cell as a result of the fuel cell reactions, and heat is also produced as a result of the process that can be used to heat the home. Hydrogen fuel cell boilers are home units that utilise the heat from the hydrogen fuel cell for hot water and heating, while also generating electricity for the home. Boilers that generate heat and electricity are referred to as Combined Heat and Power (CHP) units. Although it should be noted that CHP units commonly use combustion engines - fuel cell technology uses a chemical process rather than the burning of fossil fuels, and thus offer very little pollution in comparison to the combustion counterpart. Just like a conventional gas boiler, most micro CHP heating systems are powered by natural gas. The manufacturer Viessmann in partnership with Panasonic, have developed the Vitovalor PT2 hydrogen fuel cell boiler. Viessmann has a base in Telford, Shropshire and is a German manufacturer of heating, industrial, and refrigeration systems headquartered in Allendorf (Eder), Germany. Similar to a traditional boiler, the Vitovalor PT2 takes natural gas from the grid to heat water, but also uses fuel cell technology to extract hydrogen from the gas and convert it to useable energy. Viessmann claim the Vitovalor PT2 is optimised for maximum runtimes, thereby creating great potential for reducing electricity bills. The unit has an electrical output of 750 W and a heating output of 0.9 kW, together with the gas condensing boiler that starts on demand to cover peak loads, a total heating output of 11.4 kW,
19.0 kW, 24.5 kW or 30.8 kW is available, the system can generate electricity for 45.5 hours without interruption until the fuel cell needs a 2.5 hour break to regenerate; any excess electricity can be exported to the grid. This CHP solution generates electricity and useful heat simultaneously, offering high efficiency that can help to reduce energy bills and cut carbon footprints. Similar in size to a standard domestic boiler, the CHP system is

ideal for use in a range of environments, including detached or semi-detached homes and small commercial buildings. At the time of writing the Vitovalor PT2 is considerably more expensive than a conventional gas boiler, potentially costing between £10,000 and £15,000, but the cost includes installation. It has a guaranteed life of over at least 80,000 hours. The Vitovalor won two awards in 2018: The fuel cell mCHP appliance won the 2018 Innovation of the Year: Technology – Physical, and also Home Energy Product of the Year, at the Energy Awards ceremony in London. Fuel cell heating systems have been used in living spaces in Japan since 2009, with over 200,000 such systems now in use there.

Liquid air energy

During November, 2020 it was reported that work was commencing on possibly the world's first major plant to store energy in the form of liquid air near Manchester in England. Fundamentally it is the exploitation of cryogenics which relates to, or involving the branch of physics that deals with the production and effects of very low temperatures to generate electricity. Fundamentally the concept is to use surplus electricity from the Grid to compress air until it becomes a liquid at -196 degrees Celsius. At times of peak demand the liquid air will be warmed so it expands and the resulting rush of air will drive a turbine to generate electricity, which then can be exported back to the Grid – working in the same sense as a Pumped Storage Scheme such as Dinorwig in North Wales, which pumps water to a high level during off-peak, to be released at times of peak demand to drive a turbine, which in turn drives a generator to produce electricity, noting in this case water is used and not air. The project will be the largest in the world outside of pumped hydro schemes, which require a mountain reservoir to store water, which by their very nature are limited in the UK. The Manchester project can also be defined as a Liquid Air Battery or Liquid Air Energy Storage (LAES) and is being developed by Highview Power and is due to be operational in 2022. It is claimed the 50 MW facility at the Trafford Energy Park, Manchester will store enough power for roughly 50,000 homes for five hours. The system was devised by Peter Dearman, a self-taught backyard inventor from Hertfordshire, and it has a £10 million grant from the UK Government. The Highview LAES will store 250 megawatt-hours of energy, almost double the amount stored by the largest chemical battery built by Tesla in South Australia. Highview is developing other sites in the UK, continental Europe, and the U.S., although the Manchester project will be the first. LAES has the potential to fill a gap in the market with medium-to-long duration electricity storage technology. Dearman claims his invention can be 60 per cent to 70 per cent efficient, depending how it is used. Although less efficient than batteries, the advantage of liquid air is the low cost of the storage tanks, enabling them to be easily be scaled up. Also, unlike batteries, liquid air storage does not create a demand for minerals which may become increasingly scarce as the world moves towards power systems based on variable renewable electricity. Mr Dearman opined that batteries are really great for short-term storage, but they are too expensive to do long-term energy storage – that is where liquid air comes in. Prof John Loughhead, from the Government's Business and Energy Department, has previously praised the technology.

Biomass power stations and domestic boilers

It is claimed that biomass power stations and domestic boilers are *green* as they burn wood pellets, chips, or logs to produce hot water for heating and domestic use. But where is the saving in greenhouse gases and security of supply when a power station has to import the majority of the necessary fuel from foreign countries. During 2018 over 80 per cent of wood pellets were imported from the United States and Canada. The sourcing of large quantities of biomass in the form of wood pellets from the USA and Canada is clearly environmentally challenging due to the carbon emissions associated with its transportation over thousands of miles in dirty diesel powered shipping and land transport, not to overlook the impact on land use in these countries. It should also be noted that trees have a considerable growth period. Surely it is indefensible to pay almost £1 billion a year in subsidies to firms for burning wood in power stations. Drax power station in Yorkshire burns vast amounts of imported wood chips. Drax was once Britain's biggest coal-fired power station. It now burns millions of tons of wood pellets each year – chips that do absolutely nothing in mitigating atmospheric pollution – in fact they are far more polluting than coal, and to compound the felony the use of wood chips encourages deforestation.

Regarding biomass domestic boilers it should be noted that they are significantly bigger than traditional boilers due to the nature of the fuel - additionally there is the requirement to store the fuel on the property, which has to be replenished regularly, and manually fed into the boiler. These boilers also require a hot water cylinder and need regular cleaning maintenance.

Great leaps in technology

To suggest the possibility of electricity from space will surely conjure up thoughts of science fiction. But since the end of the Second World War the advances in science and technology have been mind boggling. What was once science fiction is now science fact! As a small boy in the 1940s such things as coloured TV beamed from satellites in space, smart phones, hand held electronic calculators and the Internet, would have been the stuff of comics and space heroes such as Dan Dare and Flash Gordon. Back then to even consider that large amounts of electrical energy could be the subject of transmission without the employment of suitable wires and cables would have been deemed ridiculous. But not so now, with the exponential growth in science and technology it is now a viable proposition, such are the advances that have been made since the last World War. Indeed modern living standards would have been beyond the wildest dreams of ordinary folk in the 1940s and 1950s. Most houses did not have the convenience of central heating and double-glazing, but were warmed by a coal fire, with many houses in the towns having coal-gas lighting. Houses did not have bathrooms, let alone the luxury of en-suites, but had outside toilets. Imagine these days living in a house without hot water taps or inside toilet. I was brought up in a house with only a single cold water tap in the kitchen and an outside WC; there was an oven above the coal fire in the living room - although the house did have both coal-gas and electric lighting – very handy when the power failed as it did quite often in those days!

It is truly amazing how technology, especially communication, has advanced ever since Alexander Graham Bell patented the telephone on March 7, 1876. It is indeed truly astounding the advances that have been made in electrical communication and electronics over the last 70 years. I can still remember, as a young boy, constructing crystal sets with copper wire, a spent toilet roll and germanium diode; the 1N34A germanium diode was used in most crystal radio sets. The enamelled covered copper wire was wound around the toilet roll and the windings (tuning coil) were secured with balsa cement (glue used in making balsa wood and tissue paper model aircraft). A narrow band of bare copper wire was created by scrapping the enamel off the copper windings and stations were tuned by moving a wire across the coil. This wire was suitably connected to the wiring circuit – a set of high impedance headphones provided the sound as there was not enough energy to operate a loudspeaker. It should be noted that early crystal sets did not use germanium diodes but employed what was commonly known as a cat whisker detector, which consisted of a piece of crystalline mineral, usually galena, with a fine wire touching its surface. Unamplified radio receivers that used crystal detectors were called crystal radios. The crystal radio was the first type of radio receiver that was used by the general public, and became the most common type of radio until the 1920s. At a later stage, and as a hobby, I progressed to building one valve, Tuned Radio Frequency (TRF) radio sets or wireless, as it was known in those days. Although the main radio in the house was a superheterodyne receiver, commonly known as a superhet wireless, which

meant you could tune directly to a radio station, unlike the early TRF version of a wireless. For the technically minded a tuned radio frequency wireless was a type of radio receiver that is composed of one or more tuned radio frequency (RF) amplifier stages. These stages amplify the signal of the desired station to a level sufficient to drive the detector, while rejecting all other signals picked up by the aerial. This was followed by a detector (demodulator) circuit to extract the audio signal and usually an audio frequency amplifier. This type of receiver was popular in the 1920s. Early examples could be tedious to operate because when tuning in a station each stage had to be individually adjusted to the station's frequency, but later models had ganged tuning, the tuning mechanisms of all stages being linked together, and operated by one control knob. By the mid 1930s, it was replaced by what was known as the superheterodyne receiver. To receive a good signal and be able to tune in many stations in those days meant having a good external aerial and earth connection to the set. Again for the technically minded, The superheterodyne receiver, often shortened to superhet, is a type of radio receiver that uses frequency mixing to convert a received signal to a fixed intermediate frequency (IF) which can be more conveniently processed than the original carrier frequency. Virtually all modern radio receivers use the superheterodyne principle. Back in July 1945 the BBC started broadcasting the Light Programme, which consisted of light music and entertainment, and is now Radio 2. The other main national broadcast programme was the Home Service, which mainly covered news and current events, and is now Radio
4. A further national broadcast was the Third Programme, which covered classical music, and is now Radio 3. Stations such as Radios 1, 2, 3, and 4 each came into existence during September 1967. These days there are numerous radio stations to choose from – perhaps far too many?

The advent of terrestrial black and white television seemed almost miraculous to us small boys back then – to be able to watch cowboy films and such like on a small screen in your own home - although not fully replacing the magic and atmosphere of the cinema, it was still absolutely wonderful. When in the late 1960s colour television began in Britain that brought another dimension to viewers delight. In my lifetime I have witnessed germanium crystals being replaced by the thermionic valve, and in turn being replaced by transistors, leading to the marvel of Integrated Circuits (ICs). An integrated circuit, sometimes called a chip or microchip, is a semiconductor wafer on which thousands or millions of tiny resistors, capacitors, and transistors are fabricated. An IC can function as an amplifier, oscillator, timer, counter, computer memory, or microprocessor. Indeed the IC is a combination of many circuits on a semiconductor material (silicon), by integrating many transistors, pathways, logic gates, and other such components to create a particular function or series of functions. IC's opened the door for Small Scale Integration (SSI) to Ultra Large Scale Integration (ULSI), the process of integrating or embedding millions of transistors on a single silicon semiconductor microchip. This explosion in technology has led not only to the computer age, but that of satellite communication, employing electromagnetic energy transmitted in the form of radio and microwaves. Basically a communications satellite is a self-contained system with the ability to receive signals from the ground and to retransmit the signals back to the ground. During September, 2021 there were 7,941 satellites in Earth orbit with communications satellites being used by both private and government organisation. Many satellites are in geostationary orbit, which is 35,785 km, (22,236 miles) above the ground. All signals for television, telephone or internet are converted into radio signals and then sent towards the satellite via a ground transmitting satellite dish. Noting that mobile phones communicate to a mobile cell tower using radio waves, and the towers communicate with satellites using microwaves as such waves can pass through the atmosphere. The satellite dish works in the same manner as the reflector for a torch or car headlights. A signal is produced or reflected from a focal point. This signal reflects off the satellite dish and travels towards the satellite. First trans- Atlantic satellite television transmission occurred on 2nd May 1965. The transmission came from the USA via the geosynchronous satellite Intelsat I, nicknamed 'Early Bird'. Then on the 3rd March 1966, Phase Alternating Line (PAL) colour television system was officially adopted for the UK, and on 1st July 1967, regular colour transmissions began on BBC2.

Satellite television

Sky Television plc was a public limited company which operated a nine-channel satellite television service, launched by Rupert Murdoch's News

International on 5th February 1989. Sky Television and its rival British Satellite Broadcasting suffered large financial losses and merged on 2 November 1990 to form British Sky Broadcasting (BSkyB). At the time of writing the full provision of broadband across the whole of the UK is still to be achieved. It remains problematic in remote and rural areas where cabling is geographically challenging and not economically viable. Currently there are about 1 million customers connected to satellite services, and it would appear that satellite broadband will be the answer. Unfortunately current costs inhibit many people for opting for the satellite choice – but the arrival of OneWeb and SpaceX's new array of satellites may prove the answer. Satellite television basically transmits electromagnetic energy from the ground to a satellite in Earth orbit, which in turn transmits the electromagnetic energy back towards the ground. Looking to the future for space communication OneWeb whose headquarter based in London, is a global communications company that has the capability to deliver broadband satellite internet services to a geographically global customer. The company is building a network of 650 Low earth orbit (Leo) satellites designed to create a global fast broadband service for remote areas. During December, 2020, OneWeb launched 36 satellites from a site in Russia to give a total of 110 in orbit, and 650 satellites will be built - with the target of creating up to 6,000 satellites.

This is not to overlook SpaceX's new array of Starlink communication satellites. On May 23rd, 2019, SpaceX launched 60 Starlink satellites into orbit on a Falcon 9 rocket from Cape Canaveral Air Force Station in Florida. The satellites are the first of a planned 12,000-satellite mega- constellation to provide Internet access to people on the ground. These satellites, orbit at approximately 440 kilometres (273 miles) above the Earth, putting on a spectacular show for ground observers as they move across the night sky. A glittering line of lights was visible in California's night sky on May 7th, 2021, as dozens of SpaceX Starlink satellites formed a 'satellite train' in space - part of a vast network of satellites that Elon Musk's SpaceX intends to use for its Starlink internet service. Despite being relatively small, the satellites can look bright as solar light bounces off them to observers on the ground; they were also visible in the night sky over many other parts of the USA. While the Internet community will obviously benefit, on the downside astronomers are concerned about over potential interference with astronomical observations. With thousands of such mini satellites and depending on orbital inclination, astrophotography could be completely ruined after dusk and before dawn as the satellites stream across the sky progressively spreading along their orbit. Some scientists have already expressed concern about the sheer number of bright satellites in the night sky.

That number will swell, as companies such as OneWeb in addition with Amazon and Telesat, are planning mega-constellations of their own. There is little question that the communication industry has capitalised on the use of space in the sense of sending electromagnetic energy in the form of radio, TV and Internet information from orbiting satellites to the ground.

Electricity from space

But what of the prospect of sending ELECTRICITY from space – that is, electrical energy that can be of such magnitude that it can be used in the home for lighting, cooking and heating et cetera. This should not come as a surprise with the advances in technology which now offer a realistic exploitation of that great source of energy in the sky that we call the Sun - a source of limitless energy. Currently we only exploit a very, very small fraction of this energy by employing solar panels to generate electricity. Terrestrial solar panels derive their electricity from space by converting the rays of the Sun into electricity - solar radiation is energy, and when this energy hits a solar panel it reacts with silicon crystals in each solar panel to produce an electric current. This electric current can then be used locally in a home or business or fed into the National Grid as explained in this book. It should also be recognised that space engineers include solar panels on satellites, and indeed on the International Space Station (ISS) to produce the energy needed to run electrical circuits. This requirement is necessary as it is not possible to provide further fuel to the satellites after they have launched - free energy from the Sun will satisfy this need. In 1941 the science fiction writer, Isaac Asimov, published a short science fiction story called 'Reason' in which a space station transmits energy collected from the Sun to various planets using microwave beams. But dear reader, instead of a space Station and beaming microwaves to various planets, why not satellites in Earth orbit beaming microwaves back to the ground to produce electricity.

Space Based Solar Power Systems (SBSPS) are large structures in space that convert solar energy into a form of energy that is transmitted wirelessly, and which is termed Wireless Power Transmission (WPT), to any remote receiver station. This receiver could either be on the surface of the Earth, or on a high altitude platform such as an aeroplane, or other space vehicle. Although expensive, it is the cleanest source of renewable energy that has the capacity to provide more energy than the world consumes or is predicted to consume in the future. Basically solar cells are needed to convert sunlight into electricity. This electricity is then converted into microwaves, and beamed to a special receiving antenna, known as a 'rectenna' on the ground. The rectenna would turn the microwaves back into electricity, and then wires would carry it to the local electricity network. A rectenna is basically a rectifying antenna, that is, a special type of receiving antenna for converting electromagnetic energy into direct current electricity. They are used in wireless power transmission systems that, as mentioned above, transmit power by radio waves. For the technically minded a simple rectenna element consists of a dipole antenna with an RF diode connected across the dipole elements. The diode rectifies the alternating current induced in the antenna by the microwaves, to produce direct current, which then powers a load connected across the diode. Schottky diodes are commonly used as they have the lowest voltage drop and highest speed and therefore have the lowest power losses due to conduction and switching. A large rectenna will consist of an array of many such dipole elements.

Therefore, dear reader, you will not be surprised that it came as a great shock to discover that during November, 2020 the Government, with its track record on wasting billions on large scale wind and solar farms, has actually commissioned new research into Space Based Solar Power Systems. It would appear thank heaven (pun intended), that there are cerebral forces at work in the energy and associated government departments after all. A giant solar power station in SPACE generating energy for the UK and matching the output of a nuclear plant is brilliant news – such satellites would harness the rays of the Sun and unlike ground based solar panels, orbital satellites would harness solar rays without drawbacks of weather or darkness at night. Indeed, the Government has ordered research into a multi-million pound solar satellite called CASSIOPeiA which stands for Constant Aperture, Solid-State, Integrated, Orbital Phased Array. This concept has the capability to augment and eventually replace traditional terrestrial power stations - the exception being the future introduction of nuclear fusion power stations. The fundamental challenge until now has been to engineer a design with an economic power-to-mass ratio that can be launched into geo-synchronous orbit to collect sunlight and efficiently convert it to a safe beam of microwaves, which is directed to a terrestrial rectifying antenna (rectenna) and feeding into the local power grid. CASSIOPeiA will be a mile long and weigh 2,000 tons and expected to cost between £10 billion and £15 billion potentially generating in the region of 2000 MW (2 GW) of power, which is equivalent to a large gas-fired (CCGT) or nuclear power station. Eventually five such satellites could be sent to a fixed point over the UK. The estimate for Hinkley Point at the time of writing is about £20 billion. Thus a space solar project would cost about half of that for a nuclear fission power station, and without all the terrifying short-comings of fission nuclear plants – go figure!

For decades NASA scientists & aerospace engineers from the world's top institutions have been researching a commercially viable design for a space-based solar power station, which surely is the last word in renewable, sustainable power generation apart, of course, from Nuclear Fusion. Solar power satellites should now no longer be seen as requiring unimaginably large investments. Space solar power systems possess many significant environmental advantages when compared to other alternatives.

Sensible UK energy policy

In this chapter we have seen a number of options to a better and more secure way of producing and using energy based on our current technical knowledge. Thus until a truly GREEN sustainable and economical means of producing electrical energy in the UK that can be implemented, it is the author's opinion that at this moment in time the Government strategy should be based predominantly on FRACKING and the wide use of efficient, relatively clean and cheap to build, natural gas power stations of the CCGT variety. As an island nation which experiences the second highest tidal range on the planet, then natural gas-fired power stations should be complemented by tidal and ocean current electricity generation. It should be understood that natural gas is far less damaging to the environment than oil or coal, producing 60 per cent less carbon dioxide.

Other means of GREEN generation such as hydro should also be added to the mix. Interconnectors have the potential provide a useful role if foreign electricity proves *cheaper* at the time, thus saving both UK natural resources and emissions, especially if the import is Norwegian environmentally friendly hydro-power , and there is also the option to sell other countries, as an when, electricity at a profit. Most *importantly* this strategy would offer security of supply and enable the UK not to be dependent on foreign countries for its vital requirement of electricity – a win, win scenario – and, in my opinion, I have yet to come across a better option. Of course, the ultimate prize is the energy of the Sun – **FUSION POWER STATIONS** - the Holy Grail of electricity generation.

The advent of nuclear fusion power stations will effectively bring the Sun to us, which will negate all other means of so-called green electricity generation, including energy from space satellites as *ground based* fusion power stations are more secure and easier to maintain than those in Earth orbit. Critics may attempt to ridicule the possibility of nuclear fusion power stations, but historically critics said heavier-than-air machines would never fly – but try telling a stealth bomber pilot that, or indeed that of a 2000+mph, Lockheed SR-71 Blackbird, jet engine aircraft – I will refrain from mentioning astronauts and the Moon landing as that would be a step too far (pun not intended)!

I do not wish to be a prophet of doom, but simply acknowledge the fact that if the Earth should experience an enormous SOLAR STORM as mentioned earlier in the book - then all bets are off as they say, as the country will be thrust back into the Dark Ages. The possibility of *more* than one extreme CME event is also on the cards, and we can only pray that World Governments are taking steps NOW to minimise the damage from any such event – I dread to think the consequences if they are not – so we will end this chapter by looking more closely at this phenomena and the past history of known significant events.

Space weather

A book on future electrical generation would, in my opinion, be irresponsible without discussing and recognising *space weather* and that a very powerful solar storm could render all such dialogue academic. Without doubt our highly technological society is dependent on a secure electricity supply – electricity is now the life blood of the country - but with all things in life it is not without its problems. In the case of electricity distribution and usage there is an 'Achilles Heel' in that electrical systems are highly susceptible and very vulnerable to intense Solar Storms. Approximately every 11 years the Sun begins a new sunspot cycle; we are now entering Solar Cycle 25 which began in December, 2019. The numbering system for Solar Cycles goes back to 1755 when the recording of solar sunspot activity began. The previous Solar Cycle 24 began in December, 2008 where activity was minimal until early 2010.

New solar cycles tend to mark periods of violent eruptions and magnetic explosions creating solar storms that send deadly radiation out into space.

As briefly mentioned in Chapter One, large solar storms are known as Coronal Mass Ejections (CMEs) being large expulsions of plasma and magnetic field from the Sun's corona. CMEs are created from a solar storm resulting in a massive burst of solar wind and magnetic fields which rise above the solar corona and are released into space. Most ejections originate from active regions on the Sun's surface, such as groupings of sunspots associated with frequent flares. Solar maximum (solar maxima) is a period of greatest solar activity in the 11 year solar sunspot cycle. Luckily these storms are a rare occurrence, but when they do happen, apart from satellites, they can also threaten activities on the Earth and can be very hazardous for life on the planet, Scientists fear that the Earth could one day be hit by a major disruptive CME, that could do lasting and substantial damage to the extent of a collapse in our high technological society, possibly resulting in a minor extinction scenario. A 2019 study has shown the Sun is capable of discharging, in quick succession, a number of CMEs, with just one, as already mentioned, having the potential of wiping out technological equipment and networks, causing chaos across the planet. CMEs consists of a billion tons of matter travelling at a million miles an hour through space, and although these storms are huge and powerful, they are very tenuous and widely dispersed with just a few particles per cubic centimetre such that much of the power they have to affect us comes from their magnetic fields resulting in geomagnetic storms. With current space technology and the Sun being 150 million kilometres (93 million miles) from the Earth, we do have some warning and time thus enabling limited precautions to be taken. Although it takes light only 8 minutes to make the journey to Earth, it usually takes a CME two to four days - extremely fast ones have been known to reach here in just over a day. We are lucky in that our understanding of the mechanisms behind solar storms and solar energetic particles is likely to advance quickly over coming years thanks to data that will be gained from two spacecraft: ESA's Solar Orbiter and the NASA Parker Solar Probe which were launched in February 2020, and August, 2018, respectively. Until a full network of surveillance satellites and means to protect satellites, ground networks and equipment can be delivered, the continuance of life in our modern high technological cities and towns is at peril. A catastrophic CME could bring about the collapse of modern technological societies, with possibly only indigenous tribes on the planet surviving. It is wise to remember that although indigenous societies make up only 5 per cent of total global population, they survive in close-knit families, having a great respect for the wonders of nature, surviving by hunting, gathering plants and growing crops. During 2020 astronomers using NASA's Transiting Exoplanet Survey Satellite (TESS) and Kyoto University's SEIMEI Telescope, observed a young sun-like star called EK Draconis for 32 nights. These observations resulted in spotting a titanic superflare on the young star and an eruption of blazing-hot plasma amounting to quadrillions of tons of electrically charged particles, a blast 10 times more powerful than any seen before by astronomers from a Sun-like star. This powerful CME is a sobering reminder that our Sun may have generated such eruptions in the past, blasting the Earth with high-energy radiation, and may do so again in the future. Such CMEs are thought to be relatively infrequent in stars types such as the Sun, but it is wise to remember that even more modest storms can damage satellites, play havoc with power grids, and can have a serious impact on Earth and human society. It is thought that powerful CMEs may have been much more common during the early years of the Solar System, having a major role in the evolution of planets like Earth and Mars. Indeed, the Martian atmosphere is very thin compared to the Earth's, although astronomers think Mars had a much thicker atmosphere in the past and CMEs may help to understand what happened to Mars over billions of years. If you find the observations relating to CMEs a bit fanciful then the following *verifiable events* should change your mind.

1859 Carrington event

During 1859, a large solar storm called the Carrington Event caused widespread problems with telegraph systems across Europe and the United States. Given our reliance on electricity today, a repeat storm of such magnitude could be far more devastating. Large solar flares often occur during a solar maximum, indeed the solar storm of September 1859, (during Solar Cycle 10), the largest recorded geomagnetic perturbation, and known as the Carrington Storm after the English amateur astronomer Richard Carrington (1826-1875), struck the Earth with such intensity that the northern lights (aurora borealis) could be seen as far south as Rome. This 1859 storm took down parts of the recently-created U.S. telegraph network, starting fires, shocking some telegraph operators. In today's modern society CMEs along with solar flares of other origin will disrupt radio transmissions, and cause damage to spacecraft electronics, and increase drag on satellites so they consume more fuel to maintain their proper orbits. But the most serious potential for damage are electrical transmission lines, resulting in potentially massive and long-lasting power outages. This is not to overlook the Internet and the problems outages would cause. Solar storms, and particularly CMEs, can pose a health threat to astronauts in space or airline passengers passing over the poles, where protection from Earth's magnetic field is at its weakest. The most worrying aspect of CMEs is that the 1859 solar storm observed by the astronomer Richard Carrington begs the question of when will the next big event take place?

1921 New York Railroad Storm

How many readers are aware of a lesser known geomagnetic storm which occurred on 13th and 15th, May, 1921 and was part of solar cycle 15. This storm triggered a number of fires worldwide, including one near Grand Central Station (Grand Central) and thus it became known as the '1921 New York Railroad Storm'. The storm was extensively reported in New York, which was a centre of telegraph activity due to being a railroad hub. Additionally, telegraph services across the U.S. first slowed and then virtually stopped at about midnight on 14th May due to blown fuses and damaged equipment. Radio propagation was enhanced during the storm due to the effect on the ionosphere resulting in unusually good long- distance radio transmission/reception. It was reported that undersea telegraph cables were also affected by the storm. The damage was not confined to the USA as problems to telegraph systems were also reported in Europe and the Southern Hemisphere. In Sweden the Karlstad telegraph and telephone exchange was 'burned out', and the storm interfered with telephone, telegraph and cable traffic over most of Europe. There were reports of extensive damage to electrical equipment, including electrical, telegraph and telephone wires in various parts of the globe such as India, New Zealand and the UK.

1967 Polar surveillance radar blackout

Disturbingly during the Cold War with Russia, a blackout of U.S. military polar surveillance radars during May, 1967 led to the scramble for nuclear war until it was realised (thankfully) the cause was due to a solar storm.

1972 Solar storms

The solar storms of August 1972 were a historically powerful series of storms; with intense and extreme solar flare, solar particle and geomagnetic storms in early August 1972. The storms resulted in damage to several satellites rendering one permanently inoperable. On land there was disruption to communication and power networks. Long forgotten and not widely known, as a result of these magnetic storms (and near the end of the Vietnam War), was the accidental detonation of numerous U.S. naval sea mines on 4th August, 1972 near Haiphong, North Vietnam.

1989 Hydro-Quebec power loss

In March, 1989 a solar storm (during Solar Cycle 22) caused the Hydro-Quebec (Canada) power grid to go down, and the resulting damage and loss in revenue were estimated to be in the hundreds of millions of dollars. It should be noted this solar storm was approximately a third less powerful than the Carrington event, and still resulted in the collapse of the Hydro-Quebec electricity transmission system. Due to the very long transmission lines of the Hydro-Quebec power grid, coupled with the fact that most of Quebec sits on a large rock shield (high ground resistivity), these factors inhibited current flowing to ground (earth) and as a result the electrical current finding a less resistance path along the 735 kV power line, which as a consequence tripped out circuit breakers. This resulted in more than six million consumers losing their power for up to nine hours. Much further afield a number of polar orbital satellites lost control for several hours and the GOES weather satellite had communication problems resulting in the loss of weather images.

1994 solar storm

In 1994 a solar storm caused major malfunctions to two communications satellites, disrupting newspaper, network television and nationwide radio service throughout Canada. Other such storms have affected systems ranging from mobile telephone service and TV signals to GPS systems and electrical power grids.

2000 Bastille solar storm

The July 14th, 2000, solar storm was named the Bastille Day Solar Storm after the storming of the Bastille, Paris, France during the eighteenth century on July 14th, 1789. The storm resulted in a number of radio blackouts and some satellite damage. As a result of solar-monitoring satellites, the storm proved significant for understanding of weather on the Sun, helping scientists in formulating a general theory of how eruptions occur, and protecting the Earth from future large events.

2003 Halloween solar storm

During August 2003, Canada, as well as the northeast United States, a solar storm caused numerous blackouts, and jammed shortwave radio frequencies used by commercial pilots. This prompted some observers to speculate that the Kremlin was jamming radio signals; a number of satellites actually lost control for several hours. Later that year a series of storms occurred from mid-October to early November peaking around 28th and 29th October causing failure of power grids. The Advanced Composition Explorer (ACE) satellite was damaged and the Solar and Heliospheric Observatory (SOHO) temporary failed. These series of storms became known as the Halloween solar storm of 2003.

2012 Solar storm

Arguably a very lucky escape when NASA's spacecraft, the STEREO (Solar TErrestrial RElations Observatory), detected a coronal mass ejection on the Sun on July 22nd, 2012. Had this eruption occurred nine days earlier, it would have hit the Earth, with the potential to wreak havoc with the electrical grid, disabling satellites effecting GPS, and disrupting our increasingly electronic lives - the solar bursts would have enveloped Earth in magnetic fireworks matching the largest magnetic storm ever reported on Earth–namely the aforementioned Carrington event of 1859 – a recent study estimated that the cost of a solar storm like the Carrington Event could reach £1.88 trillion worldwide. It has been suggested that there is about a 10 per cent chance of another intense storm in any particular year - no one really knows how bad a storm such as the 1859 event could be – remember before the 19th century there were no electrical or internet networks across the planet. To reiterate, as it is vitally important, that today we are totally dependent on our power and communication networks. Everything from banking to buying groceries depends on electrical energy and the internet in its many form of usage – imagine the sudden shock to society if the electrical power *suddenly failed* coupled with television and radio transmission, telephone landlines and the Internet going down – we are so reliant on electronic transactions for everything - everyday items such as fuel for the car would become unobtainable, railways dependent on electrical power would fail, people would be unable to get cash from the bank, satellite communication and TV would fail.

2021 storms

At the time of writing (October, 2021) a solar storm caused moderate problems manifesting itself in the Aura Borealis being visible as far south as New York and also in Scotland. A number of Internet sites went down and there was disruption of ITV in the UK. The storm was rated at a G2 geomagnetic solar storm on a scale of G1 to G5, with G1 defined as minor, G2 moderate, G3 strong, G4 severe, and G5 the strongest. A period of increased solar activity is forecast as we enter Solar Cycle 25, with a peak of 115 sunspots being predicted during 2025.

The above solar storms are far from being comprehensive as solar storms are numerous and have been recorded since 1582, but the examples mentioned sufficiently indicate the potential threat of CMEs to 21st Century technological societies.

On June 10th, 2021 Scott McIntosh, a solar physicist at the National Centre for Atmospheric Research (NCAR), said something big may be about to happen on the Sun, which he and his colleague Bob Leaman of the University of Maryland, Baltimore County, USA, have called a 'termination event'. This new idea according to the two scientists is that vast bands of magnetism are drifting across the surface of the Sun. When oppositely-charged bands collide at the equator, they annihilate (or 'terminate'). The resultant magnetism from the collision can kick-start the solar cycle into a higher gear. According to Scott McIntosh an early Termination Event would mean the new Solar Cycle 25 could have a magnitude that rivals the top few since record-keeping began. It should be pointed out though, that most solar physicists believe Solar Cycle 25 will be weak, similar to Solar Cycle 24 which barely peaked back in 2012- 2013, and orthodox models of the Sun's inner magnetic dynamo favour a weak cycle and do not even include the concept of 'terminators'. But one thing for sure is that scientists still have a lot to learn about the Sun and can only predict *how serious* the situation would be should the Earth experience a storm equal to, or indeed, more powerful than the Carrington event.

Finally, it is interesting to note that human bodies are also connected to the Earth's magnetic field, and an electromagnetic aura surrounds the body. Indeed, the brain is an electro-chemical organ, and the nervous system is a complex collection of nerves and specialised cells known as neurons that transmit signals between different parts of the body – the system coordinates actions of the body in response to changes in the environment
- in effect, it is essentially the body's electrical wiring. In 2014, a scientific study suggested that the risk of stroke increases during geomagnetic storms, and there are other potential effects of a powerful solar storm on the human body, which are outside the scope of this book.

CHAPTER NINE

Qui audit adipiscitur

Who Dares Wins

WILL DC DOMINATE

During the early days of the electricity industry a crossroad was reached, offering the choice between a DC path and an AC path. As explained earlier in the book the AC path was taken as it offered less power loss in long cables, especially in the transmission network. It is interesting that we have possibly reached this same crossroad again due to the advent of HVDC and the 21st century multitude employment of electronic devices that demand DC. The concept of the potentially fatal 230 volt AC supply to homes surely also needs appraising in the light of modern advances in technology. Please note that the voltage and tolerance for an electricity supply in the UK is 230 volts -6 per cent +10 per cent. This allows a voltage range of 216.2 volts to 253.0 volts. Nowadays home electricity requirements do not need such a high voltage and domestic devices will work with DC. How many readers appreciate that radios, television, computers, mobile telephones cannot work on AC as they are dependent on electronic circuitry, which requires DC, thus having to employ rectifying components to change AC to DC, plus components to lower the level of the voltage – a computer, for example, will function quite happily on twelve volts - laptops use re-chargeable batteries rated at 10.8 volts. To charge a battery from the mains necessitates a device (charger) that reduces and rectifies the AC mains to the required DC level. It can be argued that the concept of the distant AC power station is now outdated, and more localised DC generation should be the reality, with UK Government planning for a society where DC is the future. We have seen earlier in this book the historical reasons we have a potentially hazardous and relatively high AC fed to our homes - that the very first power stations generated DC - but due to power losses in the distribution cables customers had to live very close to the source of generation. With the advent of the transformer AC generation resolved the dilemma, where power could be transmitted over long distances without the losses of DC transmission. That is why the National Grid transmits up to 400 kV over long distances. It is could be said to be somewhat ironic that High Voltage *Direct Current* (HVDC) these days is used to transmit electricity over very long distances such as in sea-bed cables (Interconnectors). Most countries produce AC and a lot of power stations are sited far from towns and cities necessitating in long and expensive transmission lines. In the UK the dominant fuel used in large power stations are currently fossil-fuels, with natural gas accounting for about 33 per cent of usage. In other countries coal is still unfortunately widely used. Most of the electricity in China comes from coal, which accounted for 65 per cent of the electricity generation mix in 2019. While the UK on 18th March, 2021 at 15:20 GMT, was producing just 1 per cent of electricity from coal, amounting to 314 MW of power.

Countries such as the United Kingdom did not have the same water resources as the United States, Canada or Norway, but luckily the UK had plenty of fossil-fuel in the form of coal. So it is not surprising that Britain's early power stations were fired by coal. Thus, if it were not for the problems relating to power loss in the cables and subsequent high cost, the UK would most certainly be using DC instead of AC in homes, offices and industry. Therefore is it not time to challenge the continuing use of AC - significant costs would be saved with a lower level DC supply, as numerous items would not need the necessary additional components to convert AC to DC; a significant reduction in overall manufacturing costs. There are now AC power sockets on market that have USB ports for charging items such as mobile telephones, negating the need for a separate charger for the mobile and thus a saving in cost. At the time of writing a flush fit, electrical double switched 13 Amp power socket with two USB charging ports, rated at 3.1 Amp, 5 Volt, 15.5 Watt, will cost just under £12. Such sockets are easily fitted when changing from an existing conventional double switched socket. Roof-fitted solar panels generate DC and require what is known as an Inverter to convert the DC to AC before being connected to the house or local AC network. This device results in an additional cost to solar panel systems with the price of inverters ranging from £300 to £1200 depending on the size of the solar array. Mobile telephones, laptops, and tablet computers all need adapters to enable charging from the house sockets. And every time an adapter is used up to about 20 per cent of electricity is lost as heat energy, which is noticeable by the adapters heating up. Mobile telephones use approximately 2 to 6 watts when charging, while a charger left plugged in without a telephone attached will consume 0.1 to 0.5 of a watt. There are approximately 70 million mobile telephones in the UK (powerwatch.org.uk). Therefore 'charging heat loss' becomes much more significant in relation to energy wastage. The UK Government wishes to ban petrol and diesel cars in the near future, and if this scenario should become a reality, then there will obviously be a significant increase in demand for electricity. EVs carry batteries for their DC motors which will require charging from the AC mains, resulting in expenditure on the necessary charging units – the charging units would be much cheaper if they did not have to incorporate the necessary transformer and associated components for converting AC to DC - this could obviously be overcome with a DC mains. As a result of the increased use of energy-efficient technologies such as the use of LED lighting and the decline of UK manufacturing, especially in energy-intensive industries, it is not surprising the amount of energy used by the UK between 1998 and 2015 fell by 17 per cent. Indeed, the employment of LEDs have become widespread, found extensively in televisions, digital clocks, home lighting, streetlights, traffic signals, roadside hoardings, airport lighting and car brake lights, fog lights and interior lights, with car headlights becoming increasingly popular – and all dependent on low level voltage DC.

Domestic electricity

In considering a DC world it is helpful to realise how much electricity is used in the average household, and what are the major electricity appliances. Prior to 2014 energy used in a household in the UK could be in excess of 4000 kWh, whereas these days it can be as low as 3760 kWh per household. Larger houses tend to use more electricity, whilst mid-terraces and flats use the least electricity. A detached house uses in the region of 4150 kWh per year, a bungalow 3870 kWh, semi-detached house 3850 kWh, end of terrace house 3440 kWh, flat 2830 kWh and a mid-terrace house 2780 kWh per year. But should a dwelling use electricity for heating then the figures will obviously be very much higher. Compared to other countries, and according to the Organisation for Economic Co-operation and Development (OECD) the use of electricity is not that high, considering the U.S. uses on average 12,300 kWh each year, in Canada the average is 11,000 kWh, and in Australia it's 7,000 kWh. But remember the North America Continent has hot summers and vey cold winters, and Australia is not known for intense cold. Note also that homes in the UK are relatively small, and heating is primarily done with gas coupled with the fact that air conditioning is not widespread. According to the World Energy Forum, the average European usage is of a similar magnitude at 3,600 kWh, which is more than double that in China averaging at 1,500 kWh, although three times that in India averaging just 1,100 kWh. The lowest recorded average for electrified homes is Nepal at 320 kWh, according to World Energy Forum figures. Perhaps surprisingly Saudi Arabia uses a large amount of electricity per household at over 24,000 kWh per annum – this surely must be due to the local hot (40 degrees) climate and the demand for wide-spread use of air-conditioning. Regarding actual appliances in the average UK household cold appliances such as fridges and freezers consume the most electricity at about 630 kWh per year, lighting 600 kWh, consumer electronics 570 kWh, cooking 540 kWh, wet appliances 530 kWh, water heating 285 kWh and computing 250 kWh; giving a total of 3405 kWh per year for an average household. By magnitude this figure corroborates the figures given that a detached house uses 4150 kWh and a mid-terrace house 2780 kWh per year. Although fridges and freezers in total consume the most electricity (about 15 per cent of total usage) in a household, it is individual appliances such as electric heaters, dishwashers and washing machines that consume the most power. All these appliances can be designed to function with DC and at a much lower voltage level than 230 volts AC.

DC low voltage grid

With the advances in technology a total reassessment should now be made on how we can minimise the use of energy in the home, reducing the network voltage level, without the loss of any of the benefits. There is no reason why this cannot happen, with the use of smaller local power stations. Electrical appliances in the home can be designed to function across a range of say 12 volt to 48 volt DC – think of caravans and motor homes for example. Modern low-voltage motors can have a long life span with an occasional replacement of brushes in their motors. Indeed, 12V DC brushless motors are gradually becoming more available, which should make the low voltage appliances virtually maintenance-free. In the kitchen modern 12-volt refrigerators requiring only 500 watts are equal in operation to an icebox than requires 3 kW to function - although it is true that they are currently not cheap to buy. Washing machines can also be designed to work in the 12 to 48 volt DC range. Blenders, mixers and toasters will lend themselves to 12 volt working. Elsewhere in the house hair dryers, television and radios, computers and electronic devices can function comfortably with 12 volts. It can be argued that a dwelling fitted with roof-top solar panels complimented with batteries would have minimal reliance on a DC Grid connection – there could also be the option of going off-grid by having battery charging centres for periods when there is minimal solar radiation.

Hybrid consumer unit

A step towards a fully DC home could be the introduction of a hybrid consumer unit – especially for newly built houses.

The Incandescent bulb provided the standard lighting method in the UK for more than a century, but as a consequence of an EU directive, the Government banned the import of 100-watt bulbs from 2009, followed by a ban on 60-watt bulbs in 2011 and a full ban on all incandescent bulbs in 2012 - although it should be noted that it is not illegal to buy, sell or use incandescent light bulbs if you can get hold of them. Incandescent lighting has predominantly been replaced with Light Emitting Diodes (LEDs), and whilst LED lighting is not strictly a legal requirement, all new residential developments in the UK must include at least 75 per cent low-energy lighting, defined as lights that produce over 400 lumen at an efficiency of 45 lumen per watt of power output. Lighting uses around 20 per cent of the electricity generated in the UK and LED technology can offer up to 80 per cent saving in electricity.

LED lighting offers the opportunity for the modern consumer unit to be re-designed with two sections, one to accommodate AC for ovens, electrical showers and power sockets, and the other section for DC low level voltage circuits such as lighting. The design of the low voltage section would incorporate a small battery that is charged from the AC mains via a rectifying unit, offering a *reliable* supply for lighting in times of Grid outages. In the low voltage DC section, instead of fuses or circuit breakers, simple switches would be provided for each lighting circuit feeding the house. Each DC circuit switch could be highlighted with an appropriate LED. Such a system would greatly enhance safety, security and convenience as the house would not suddenly plunge into darkness during a mains failure at night. There would be significant saving in house cabling and component costs due to the use of low voltage, cheaper cables and switches, thus reducing overall costs and increasing safety – electrical shocks or indeed electrocution would not be possible on these DC circuits. The AC section of the consumer unit would consist of the normal circuit breakers feeding the 230 volt AC supply cables. This section of the hybrid consumer unit will still demand *respect and a safe approach* due to the AC 230 volt level. No doubt the astute reader will realise that AC power sockets currently on market incorporating USB ports, could become redundant for charging items such as mobile telephones, simply because a lighting circuit socket or adaptor would now offer a DC means of charging such devices. But has UK Government the foresight and will to set a new standard, by implementing a new UK house electrical wiring code of practice, with the ultimate target of achieving a fully DC working home? On a global scale how many readers are aware there are in excess of 600,000 homes in India alone are already powered by DC. Bangladesh today has over five million DC powered home energy systems, providing light, comfort and livelihood – and the numbers are growing.

CHAPTER TEN

'You can fool all the people some of the time, and some of the people all the time, but you cannot fool all the people all the time'

Abraham Lincoln (1809-1865)

WATER COMPANIES

Perhaps a much more apt title to this chapter should read, 'Muddying the Waters', as it is argued the privatisation of the water companies, similar to the power companies, is a perversion of true marketing forces. Marketing can best be observed in the High Street where, for example, Jack's clothing shop or stall has to compete with that of Jill's clothing shop or stall. It is not rocket science to work out that the shop or stall which offers the best service, excellent quality and competitive prices will prevail.

What marketing forces are at work in the water companies?

The reader should be fully aware that with the current legislation, household customers are NOT able to change their water supplier or sewerage service provider. The water or sewerage company that supplies your property will depend on where you live. What a surprise to find there are no 'High Street Marketing Forces' at work in the supply of water from a reservoir, river or ground source – such is the illusion and delusion! As an analogy earlier in the book it was shown that the change of Electricity Company for the consumer is purely a *paper exercise,* and the only change for the customer will be the company name on the top of the bill, and a change in the Standing and Unit Charge for that specific company; an engineer will certainly not turn up at the door to make any changes - the household will still be connected to the same supply cable, pole or underground feed, sub-station, all the way back to the Grid Network. Do not be fooled if an electricity company tells you they can directly feed you cheap Green Electricity. This can only happen if a customer has a direct link to a Green source of electricity. The electricity received at the household electricity supply meter cannot discriminate between sources of electrical energy – common sense should tell you that energy from the Grid is a mixture of various sources of generation. The household electricity supply is a good analogy to that of a privatised water company with its *machinations* and *sleight of hand,* as the water supply will arrive at the household through the same pipes and from the same source. Households that receive water extracted from a local river, such as the Teifi in west Wales, will never receive water from Lake Vyrnwy, or the Elan Valley dams as these reservoirs are *piped* to Liverpool and Birmingham respectively. The good citizens of these cities will never be fed water from the Teifi River in west Wales - how can they, unless it is taken by tanker – so where is the competition? It is basic common sense to realise that the only potable water that can honestly claim to be subject to true marketing forces is that of bottled water sold on the high street and in various supermarkets. Surely, the concept of water privatisation is nothing more than a clever but devious strategy to make money for 'Shareholders and Chief Executives' – all at the expense of the long suffering customer. The customer certainly does not have a selection of taps in their kitchen so they can decide whose water is currently the best and cheapest. The customer is still supplied from the same reservoir, aquifer, river or whatever – the supply pipes to the household have not changed only the billing - but now with separate bills for water and sewerage charges. It is worth reiterating:

*Under the current legislation, household customers are **NOT** able to change their water supplier or sewerage service provider. The water or sewerage company that supplies your property will depend on where you live.*

Research carried out during September and October, 2021, by YouGov poll, on behalf of the charity National Energy Action (NEA), which revealed that 41 per cent of households with an income of less than £20,000 bathe less to reduce their bills. A spokesperson for NEA at the time said that, "Access to safe, clean and affordable water is recognised as a human right." Indeed it is, and it is outrageous that water companies have been polluting our rivers and beaches with raw sewerage (see later in this chapter) whilst charging customers such high prices. Thus, it is well done NEA for such revelations. It was also disturbing when looking at share prices on 5th November, 2021 that Severn Trent had a price of £28.23 per share, whilst National Grid was £9.55, and Centrica just 63 pence. Other shares such as BT were £1.57 and Vodaphone £1.10. At flotation in 1989 ordinary shares in Severn Trent PLC were offered at a price of £2.40, so it would seem that Severn Trent is *'the goose that laid the golden egg'*. I do not pretend to understand the Stock Exchange, except to question how a water company with 'captive' customers (where else can they go) can be valued so much higher than another company, such as BT, Centrica or indeed the National Grid. The water industry was privatised in 1989 in England and Wales which involved the transfer of the provision of water and wastewater services from the state to the private sector, through the sale of the ten Regional Water Authorities (RWA). It should be noted that there are two forms of private sector participation in water supply and sanitation. In a full privatisation, assets are permanently sold to a private investor. In a public-private partnership, ownership of assets remains public and only certain functions are delegated to a private company for a specific period. To my knowledge, full privatisation of water supply and sewerage is an exception being limited to England, Chile and some cities in the United States. Public-Private Partnerships (PPPs) are the most common form of private sector participation in water supply and sewerage.

Unacceptable pollution

No doubt misguided supporters of water privatisation will claim that decades of underinvestment by successive governments have resulted in poor water quality, rivers were polluted, and the beaches badly affected by sewage. These have to be 'challengeable words' as surely this underinvestment is a failure of inept government and management NOT that of public ownership. It is nothing more than an attempt to muddy the waters (no pun intended).

The claims of 'cleaner rivers and beaches' under privatisation are vacuous. I speak from experience as a few years ago I suffered from an ear infection whilst swimming in the sea in west Wales. On attending my local surgery I was asked by the doctor if I had been swimming in the sea at Aberporth to which I answered, yes. I tend to have dry skin in my outer ear canal and the cracks in the skin are vulnerable to water borne bacteria. It should be noted that during July 2010 swimmers were advised to stay out of the sea at a Ceredigion beach (Aberporth) following a sewage leak. Environment Agency Wales officers who investigated at the time suggested the spill was caused by a private sewer. These days when swimming in the sea I wear earplugs and keep my head out of the water - once bitten, twice shy, as they say. It is stunning that throughout 2020 the not-for-profit water company Dŵr Cymru (Welsh Water) were responsible for more than 100,000 discharges into rivers and seas around Wales.

The contempt for the public and disingenuous nature of the water companies was clearly demonstrated and proven when Southern Water was found guilty of 51 pollution offences at Maidstone Crown Court on 11th March 2020. It was reported that Southern Water had been fined a record £90 million for deliberately dumping billions of litres of raw sewage into the sea. The company admitted an astonishing 6,971 illegal spills from 17 sites in Hampshire, Kent and West Sussex between 2010 and 2015. The offences were discovered as part of the Environment Agency's largest ever criminal investigation, which began after shellfish were found to be contaminated with E. coli. When considering what customers have to pay water companies for these services coupled with the payment to shareholders and management, it is outrageous that in the 21st Century raw sewage had been diverted away from treatment works and into the environment - disgracefully the total volume of untreated sewage dumped into the environment was estimated between 16 to 21 billion litres, or over 7,000 large-sized swimming pools. It was shocking that anglers In Bedhampton Creek, in Havant, Hampshire, reported finding sanitary towels, condoms and tissues in the water, along with a disgusting and strong smell of sewage. No doubt any fish caught were returned to the water. The offences had been aggravated by Southern Water's *'persistent pollution of the environment'* which had led to 168 previous convictions and cautions.

Unfortunately and shamefully, Southern Water are not alone, as on 22 March, 2017, Thames Water was ordered to pay fines and costs amounting to over £20 million for six separate water pollution incidents on the River Thames and its tributaries. It begs the question of how many acts of pollution are ignored or unidentified and therefore not resolved.

Enigmatically, why is it that Greta Thunberg the Swedish environmental activist and her followers not protesting outside the headquarters of water companies, when they have little problem in blocking major central London roads to the disadvantage of many good hard working folk going about their business. Do these environmentalists not regard effluent as a danger to health and well being – do the polluted rivers and **foul** coastal regions not register on their radar? Perhaps Greta Thunberg prefers the publicity and fame of the international stage challenging world leaders over climate change - how ironic that during the COP26 climate summit in November, 2021 she used **foul** language to swear at world leaders.

But how shocking and contemptible that as recently as October, 2021 a vote for a change in the Environment Bill in the Commons was voted down by Conservative MPs as the legal changes would force water companies to reduce the amount of sewerage they dump in rivers. It was only due to a backlash from campaigners and the social networking site Twitter that the Government was coerced into a climb down and introduced a new amendment for MPs to vote on. The new amendment called on the Department of Environment, Agriculture and Rural Affairs to ensure water companies secure a progressive reduction in the adverse impacts of discharges from storm overflows. But I would contend this amendment does not go far enough in addressing the countries inadequate Victorian sewerage systems - it is even more outrageous considering water companies discharged raw sewerage **400,000 times** in 2020. Additionally, film footage in October 2021 clearly showed untreated sewerage being released into Langstone Harbour, Hampshire from Southern Waters Budds Farm treatment plant. Unbelievably this is supposed to be a conservation area – why are these companies allowed such disgusting liberties in our day and age?

Since privatisation the mostly now foreign owned water companies have paid themselves in the region of £56 billion in dividends, and it does not take a degree in business management and practices, to recognise that if the profits were invested back into the water companies, instead of being paid to shareholders and management, then such health hazards can be avoided, by improving treatment plants to an acceptable and reliable standard. Although it is reassuring that the Environment Agency was very effective in bringing about the criminal prosecution of Southern Water saying it was the biggest case in its 25-year history.

IRRESPONSIBLE POLLUTION

Why is it that in our so-called technological advanced society, during periods of heavy rain, water companies are permitted to divert untreated waste water away from treatment plants, discharging sewage straight into the environment to prevent sewers backing up! This pollution should NOT be underestimated or tolerated, as it greatly increasing the chances of cholera and other diseases. Recognise that with population growth more and more people are visiting the coastal regions swimming and dipping in the sea, not to overlook contaminated rivers and lakes – we should all be lobbying our political representatives for action now! During September and October, 2021 it is a shame that Insulate Britain protesters had not glued themselves to the doors and buildings of the various water companies, rather than road surfaces. What is the point of blocking roads and stopping good citizens attending their places of employment, when the rivers and coastal regions become open sewers – have these single-minded protesters never heard of cholera and other water borne diseases – do these protesters consider a well insulated house is of greater importance than a cholera epidemic?

Dŵr Cymru (Welsh Water)

Regarding the '*trickery*' of privatisation it is revealing that Dŵr Cymru (Welsh Water) differs from other water and energy companies by not having any shareholders – they say any financial surpluses are reinvested in the business for the benefit of customers – but is this the whole truth? Dŵr Cymru (Welsh Water) serves over three million customers in most of Wales, Herefordshire and parts of Deeside. Why the *partial* privatisation of Welsh Water, and why should they be different to all the other water companies? Could it be that Wales is a devolved country having major reservoirs that feed large cities in England - the Elan Valley dams feed the City of Birmingham, the Clywedog reservoir near Llandiloes in Powys, and built by damming the Afon Clywedog in 1967 to supply water to Birmingham and the Midlands, whilst Lake Vyrnwy reservoir feeds the City of Liverpool! If Dŵr Cymru (Welsh Water) was a public company and as such owned these reservoirs situated in Wales, then it might prove challenging for English towns and cities. As a public company in a devolved country, Dŵr Cymru (Welsh Water) might feel fully justified in asking for a '*certain price*' for their water to mitigate Welsh taxpayers and assist in devolution - what price for Welsh Water under these circumstances!

Dŵr Cymru (Welsh Water) is a large employer employing over 2,500 people, looking after 92 reservoirs and maintaining 27,000 km of water pipes, managing over 800 wastewater treatment works and over 30,000 kilometres of sewers. They also have a graduate and apprentice scheme which runs annually. It is one of 10 water and sewerage companies in

England and Wales, and unlike other water companies it is a 'not-for- profit company' and has been owned by Glas Cymru since 2001. By not having any shareholders it is claimed they are able to keep bills down, help more customers who are struggling to pay their bills and also put every penny they make straight back into maintaining and improving service and protecting the environment, now, and for years to come. Dŵr Cymru (Welsh Water) claiming it is a public/private hybrid where the virtue lies in taking the best of the public and private sectors, needs challenging. For it is a shallow claim as it begs the question, as to which water company will they compete with, and much more revealing, if they claim to have found the *'perfect solution to water and sewerage'* why are the other water companies NOT following their marvellous example? In the other water companies shareholders certainly relish their dividend and the share price to increase - consider the wages of the chief executives and then examine your household bills, whilst considering the state of our rivers and coastline - who do you think, dear reader, is being taken to the cleaners!

CRAIG GOCH RESERVOIR, ELAN VALLEY, POWYS, WALES

The Consumer Council for Water, (CCWater, www.ccwater.org.uk) is a statutory consumer body for the water industry in England and Wales. They offer independent advice about water companies and if your local water company service has been poor, they may investigate your complaint. A report published by CCWater dated 13th January, 2019, called for the water firms to show more ambition in their service and investment plans for 2020 to 2025. Water firms should do more to tackle leaks, keep the taps running, control executive pay and deliver value for money. The Consumer Council was responding to the regulator Ofwat's initial assessment of companies' business plans which set out what they propose to deliver for customers and how much they want to charge for that investment. The Water Services Regulation Authority, (Ofwat, www.ofwat.gov.uk.), is the body responsible for economic regulation of the privatised water and sewerage industry in England and Wales. Ofwat is primarily responsible for setting limits on the prices charged for water and sewerage services. Ofwat commented that only Severn Trent Water, South West Water and United Utilities had offered acceptable business plans for 2020-2025. The remaining water companies needing to amend their plans, with Affinity Water, Hafren Dyfrdwy, Thames Water and Southern Water, requiring the most substantial changes. The Chief Executive of the Consumer Council for Water said that they had not seen one plan that could be considered the finished article and there was still scope for improvement across the board. He also said that many companies had not stretched themselves on issues like reducing leakage and expanding support for customers in financial hardship. Adding that most people will see their bills rise by 2025 once you add inflation and the financial rewards that companies can earn but are paid for by customers. It is important companies and Ofwat are clear with households if their bills are going to rise."

You do not need a crystal ball to foresee if water and sewerage bills will rise in the future as management and shareholders welcome profits.

A National Water Company

The Government should not only re-nationalise the water companies, but introduce a National Water Grid, and the new company could simply be called British Water. In constructing a national water grid there need not be an extensive provision of pipe-lines as a cursory glance at a map of Britain will show an extensive network of canals and many rivers, all of which could be inter-connected by minimal use of pipes and aqueducts. As an example, the River Severn rises in the Cambrian Mountains in mid- Wales, flows north to Welshpool before turning south toward Shrewsbury before entering the sea after flowing through Gloucester, at this point the River Thames is not far from the River Severn for inter-connection purposes. A National Water Grid would guarantee that every part of the mainland would not suffer from lack of water during a hot or dry summer, as there are parts of the country which will always have an abundance of water. It is no surprise that the wettest parts of the country are in the west and the north where annual rainfall is abundant with places such as Crib Coch (Gwynedd) receiving 4635 mm rainfall annually, Styhead (Cumbria) 4562 mm and Glenshiel Forest (Ross and Cromarty) 3778 mm. The driest parts of the country are, as expected, in the south-east with London having 557 mm of rain and Cambridge with 568 mm. Simply put, we have the rain, the rivers, and canals, but where is the will, imagination, and the drive? The Roman's, if alive today, would soon resolve the situation – even today the City of Rome has many public fountains thanks to the efforts of the Roman engineers.

Appendix One

Electrical units

This book has been written for the general public and as such it is realised a lot of folk will not be familiar with all the electrical units, especially those employed in large quantities. Therefore it is hoped the following explanation will be of help.

The prime reason why the power industry uses a variety of units is simply to overcome the cumbersome use of large numbers. Students of astronomy will quickly recognise the logic when measuring distances in space.

For small measurements such as the distance between the Earth and the Moon kilometres (miles) are used. The average distance being the Earth and the Moon is 384,400 kilometres (238,855 miles). Whilst the average distance to our nearest star, the Sun, is 149.597,870 kilometres (92,055,807 miles).

As we progress further out into space the numbers begin to get cumbersome and thus another unit is employed. Therefore the distance from Earth to the Sun is called an Astronomical Unit (AU), which is used to measure distances throughout the Solar System. Jupiter, for example, is
5.2 AU from the Sun. Again, when we journey further out again into space even the Astronomical Unit is too small a unit and the term light-year is used, which simply means the distance that light travels in one year. This unit of astronomical distance is 9.4607×10^{12} km (nearly 6 million, million miles). Alpha Centauri is the third-brightest star in the sky, in the constellation Centaurus, but is visible only to observers in the southern hemisphere, and is 4.34 light-years distant from the Earth.

But even the light-year is too small a unit as we delve deeper into space as the next unit is the Parsec, where the unit is about 3.26 light years (3.086×10^{13} kilometres). It is the largest unit used in astronomy and is approximately 30 trillion kilometres (19 trillion miles). The distance to the nearest galaxy, the Andromeda Galaxy is 2.54 million light-years, or 778 kilo-parsecs.

It should be noted that the 'new' British billion is 1000,000,000 (10^9). The 'old' billion was 1000,000,000,000 ($10^{12)}$, this is now the trillion.

For global communication uniformity is essential. Lack of uniformity has unfortunately been demonstrated by the loss of a space mission, when metric units were misinterpreted as 'imperial' leading to the loss of a mars mission. Indeed, In September of 1999 NASA lost a $125 million Mars Climate Orbiter because an American engineering team used English units of measurement while the agency's team used the more conventional metric system for a key spacecraft operation.

Hopefully this short excursion into the realm of astronomy and space demonstrates the method of overcoming the cumbersome use of large numbers, and the need for uniformity.

Units

In the world of electricity the basic unit is the watt:

Watt (W): The unit of power. A typical low energy bulb (LED) can use about 6 -12 watts, and a small electrical heater can use 800 watts.

Kilowatt (kW): A thousand watts. Used for domestic appliances such as white goods and electric fires.

Megawatt (MW): A million watts. Used for power generation and distribution. For example, Pembrokeshire Power Station has a capacity of 2000 MW.

Gigawatt (GW): A billion watts. Generally used to describe national demand, e.g. UK peak demand for 2019 was 47.275 GW.

Terawatt (TW): A trillion watts. Used when describing very large levels of electricity.

It is very important to recognise the above units are for INSTANTANEOUS ENERGY usage.

Where TOTAL ENERGY usage is concerned then *time* gives a different unit, such as:

Kilowatthour (kWh): Domestic electricity bills use this unit. For example, a one kilowatt (1000 watts) fire used over the period of one hour will use 1 kWh, and a two kilowatt fire will use 2 kWh. Thus an electrical appliance rated at 3000 watts (3 kW), would use 6 kWh over a period of two hours.

Megawatthour (MWh): It is cumbersome to write that the 2000 MW

Pembrokeshire Power Station has the capacity to generate 2,000,000,000 kWh over one hour, when it can be written more conveniently as 2,000 MWh.

On a larger scale **Gigawatthour (GWh)** and **Terawatthour (TWh)** are used.

SI (System International) units

Kilowatt (kW) = 1000 watts
Megawatt (MW) = 1,000,000 watts (1000 kW)
Gigawatt (GW) = 1,000,000,000 watts (1000 MW)
Terawatt (TW) = 1,000,000,000,000 watts (1000 GW)

It should be noted that 1000 watts (kW) uses lower case k, as the upper case K is the SI unit of temperature for the kelvin (K).

Unit of power usage = 1 kW consumed over 1 hour = 1 kilowatt-hour (kWh).

Indices & the Law of Indices

Indices are a useful way of more simply expressing large numbers. They also present us with many useful properties for manipulating them using what are called the Law of Indices.

The expression 2^5 is defined as follows:

$2^5 = 2 \times 2 \times 2 \times 2 \times 2 = 32$

We call "2" the **base** and "5" the **index**.

Thus the equation:

$5 \times 5 \times 5 \times 7 \times 7 \times 7 \times 10 \times 10 = 42873 \times 100 = 4,287,500$

Can simply be shown as:

$5^3 \times 7^3 \times 10^2 = 4,287,500$

Hence:

$10^2 = 100$ (Hundred)

$10^3 = 1000$ (Thousand)

$10^6 = 1,000,000$ (Million)

$10^9 = 1,000,000,000$ (Billion)

$10^{12} = 1,000,000,000,000$ (Trillion)

$10^{15} = 1,000,000,000,000,000$ (Quadrillion)

Hours in a year = 24 x 365 = 8760

Appendix Two

Wind Generation

Wind Generator Vs Coal-Fired Power Station

To determine the equivalent number of 2 MW, 123 metre (400 feet) high, wind generators required to replace a single typical coal-fired power station.

As an example the old Aberthaw B coal-fired power station had an installed capacity of 1,500 MW.

If we take the load factor of a 2 MW wind generator as typically in the order of 25 per cent, and Aberthaw B having a load factor of 62 per cent (using DTI 2005 figure), then:

Potential output of Aberthaw = 1500 x 0.62 = 930 MW
Potential output of a 2 MW wind generator = 2 x 0.25 = 0.5 MW

Thus 930/0.5 = 1,860 wind turbines.

Therefore to replace a large coal-fired power station such as Aberthaw, would require nearly two thousand 2 MW, 123 metre (400 feet) high wind generators.

Note: DTI (Department of Trade and Industry) became the Department for Business, Enterprise & Regulatory Reform (BERR) and is now The Department of Energy and Climate Change (DECC).

To determine the equivalent number of wind generators that would be needed to compete with the following seven large coal-fired power stations:

Station	Installed Capacity (MW)
Ferrybridge C	1,955
Fiddler's Ferry	1,961
Eggborough	1,960
Cottam	2,008
West Burton	1,972
Rugeley	1,006
Aberthaw B	1,500
Total	12,362

We will again take the for load factor for a coal-fired power station as 62 per cent (DTI 2005 figure), and a load factor of 25 per cent for a large 2 MW wind generator.

Total output = 12362 x 0.62 = 7,664 MW

Output of wind generator = 2 x 0.25 = 0.5 MW

Thus, 7664/0.5 = 15,328 wind generators.

Therefore to compete with seven large coal-fired power stations would require over fifteen thousand 123 metre (400 feet) tall, 2 MW wind generators.

Now 20 wind generators might extend over an area of 1 square kilometre - so 15,328 would require over 766 square kilometres of land – this would truly be utter madness!

Appendix Three

Solar Energy

Domestic Earnings 2012

The Feed-in Tariff was very generous during 2011 whereby the generation tariff paid was 54.17 pence for every kWh generated and 3.82 pence for every kWh deemed exported. This meant a 4 kW system capable of generating 4,000 kWh per annum would be paid:

(4000 x 54.17)/100 + (4000/2 x 3.82)/100 = 2166.8 + 76.4 = £2243.2

Thus over 20 years: 2243.2 x 20 = **£44,864**

Property in Pembrokeshire

The solar power array was installed on the roof of the detached garage with the panels having a tilt of approximately 35 degrees and a southerly orientation. The solar array consisted of ten 230 watt Dimplex high performance polycrystalline solar PV modules, offering a total system capacity of 2.3 kW. The capital cost of the installation was £7,818.

Solar Generation for 2012, kWh

JAN	FEB	MAR	APR	MAY	JUN	JUL	AUG	SEP	OCT	NOV	DEC	TOTAL
42.5	93.1	219.4	181.8	306.9	241.3	257.1	230.4	219.6	137.7	83.2	43.2	2056.2

During 2012 the FIT payment for solar generation was 43.3 pence per kWh generated and 3.1 pence per kWh, exported.

During 2012 a total of 2,056.2 kWh was generated.

Thus FIT payment for solar energy generation:

2056.2 x 43.3 + 2056.2/0.5 x 3.1 = 92220.57/100 = £922.2
Projecting this figure over 20 years = 20 x 922.2 = £18,444.

Pembrokeshire generation for 1 year = 2056 kWh

Pembrokeshire generation over 20 years = 20 x 2056 = 41,120 kWh

Mains electricity saved at property in Pembrokeshire

The average assessed annual mains electricity was £500 and with the fitting of the solar panels this was reduced to £294.4, saving £205.6 per annum.

Therefore total earnings per annum (solar generation plus mains saved electricity):

£922.2 + £205.6 = £1127.8

Projected over 20 years = 20 x 1127.8 = £22,556

To arrive at the true earnings the above projected earnings over 20 years necessitates the reduction cost of the capital cost of the system: £7,818.

Thus 22556 − 7818 = £14,738.

Payback time for the capital cost of the system:

7818/1127.8 = 6.9 years.

Property in Ceredigion

A total of 16 solar panels were mounted on the roof of the detached property.

The solar generation for year 2014 is used for a comparison to the property in Pembrokeshire.

Solar generation for 2014 in kWh

JAN	FEB	MAR	APR	MAY	JUN	JUL	AUG	SEP	OCT	NOV	DEC	TOTAL
88	197	331	424	471	658	567	486	424	219	166	100	4131

During 2014 the FIT payment was 14.9 pence per kWh, and 4.6 pence per kWh exported.

During 2014 a total of 4,131 kWh was generated.

Thus FIT payment for solar generation:

(4131 x 14.9)/100 + (4131 x 0.5 x 4.6)/100 = 615.52 + 95 = £711

Projecting this figure over 20 years = 20 x 711 = £14,220

Ceredigion generation for 1 year = 4,131 kWh

Ceredigion generation over 20 years = 20 x 4131 = 86,620 kWh

Mains electricity saved at property in Ceredigion

Annual mains electricity before the fitting of solar panels was historically assessed at 6,000 kWh for the property in Ceredigion.

The actual mains electricity consumed with the solar panels fitted amounted to 4,200 kWh.

At 12.28 pence per unit the cost of mains electricity before solar panels fitted was, (12.28 x 6000)/100 = £736.8

Cost of mains electricity after solar panels fitted was, (12.28 x 4200)/100 = £515.76

Thus a saving of, £736.8 - £515.76 = £221 was achieved.

Therefore total earnings per annum = £711 + £221 = £932

Projecting this figure over 20 years = 20 x £932 = £18,640

The total earnings over 20 years has to be qualified by deducting the initial capital cost of £7,100 for the solar PV system, which then gives a total earning of £11,540.

Payback time for the capital cost of the system = £7100/£932 = 7.6 years. This was deemed very acceptable both in being 'green' and as a good investment.

Generation Comparison

In comparing both properties it should be noted that Ceredigion has almost double the generating capacity as the property in Pembrokeshire and thus should generate approximately double the income that of the property in Pembrokeshire, but this has not been so.

The reason for this is predominantly due to the difference in FIT payments as the generation payment for Pembrokeshire was 43.3 pence for generation and 3.1 pence for export, compared to 14.9 pence for generation and 4.6 pence for export at Ceredigion.

If we apply the same FIT payment for Ceredigion we have:

(4131 x 43.3)/100 + (4131 x 0.5 x3.1)/100 = 1788.7 + 64 = £1,916.7

Which is almost double that of Pembrokeshire at £922.2.

Calculation shows £922.2 x 2 = £1,844.4, just £72.3 difference.

Thus we can see what a great difference the FIT payment makes increasing the Ceredigion annual solar generation from £711 to £1,916.7 a difference of £1,205.7.

With the actual rates the annual solar generation for Ceredigion being £711 compared to that (historically) of £922.2 for Pembrokeshire, a difference of £211.2.

Ceredigion solar generation (kWh)

2014

JAN	FEB	MAR	APR	MAY	JUN	JUL	AUG	SEP	OCT	NOV	DEC	TOTAL
88	197	331	424	471	658	567	486	424	219	166	100	4131

2015

JAN	FEB	MAR	APR	MAY	JUN	JUL	AUG	SEP	OCT	NOV	DEC	TOTAL
103	189	340	553	556	608	493	438	412	255	86	59	4092

2016

JAN	FEB	MAR	APR	MAY	JUN	JUL	AUG	SEP	OCT	NOV	DEC	TOTAL
85	180	356	472	580	516	457	438	312	299	165	101	3961

2017

JAN	FEB	MAR	APR	MAY	JUN	JUL	AUG	SEP	OCT	NOV	DEC	TOTAL
119	141	321	493	575	458	508	439	297	179	113	81	3723

2018

JAN	FEB	MAR	APR	MAY	JUN	JUL	AUG	SEP	OCT	NOV	DEC	TOTAL
123	212	295	422	587	636	608	413	389	266	164	41	4156

2019

JAN	FEB	MAR	APR	MAY	JUN	JUL	AUG	SEP	OCT	NOV	DEC	TOTAL
87	231	332	423	620	500	572	482	376	237	119	94	**4073**

2020

JAN	FEB	MAR	APR	MAY	JUN	JUL	AUG	SEP	OCT	NOV	DEC	TOTAL
111	182	409	527	705	462	494	434	390	202	142	84	**4142**

Over seven years AVERAGE solar generation

$$(4131 + 4092 + 3961 + 3723 + 4156 + 4073 + 4142)/7$$
$$= 28278/7$$
$$= \textbf{4,040 kWh}$$

Load factor

Electrical power generating units whether thermal power stations (fossil-fuel or nuclear), wind generator or hydroelectric generator, cannot always produce maximum output, and may produce no electricity if the unit is shut down for one reason or another.

Load Factor is a measure of the utilisation rate, or efficiency of electrical energy usage; a high load factor indicates that load is using the electric system more efficiently, whereas consumers or generators that underutilise the electric distribution will have a low load factor. The reader may come across the term, Capacity Factor. Please note it is the American electrical engineering professions synonym for Load Factor and this term is not recognised by British Standards Institution; although it is used by the British Wind Energy Association (BWEA) for reasons best known to them.

The method of determination is as follows:

Load Factor = Actual amount of electricity produced over time/Electricity that would have been produced if generator operated at maximum output 100 per cent of the same time.

Load factors are usually calculated as *per cent per year* but may be calculated for other periods of time. They must be compared only for the same base period and starting at the same point in time (e.g. January to December).

An example, using a large commercial electrical bill:

Power demand = 436 kW
Number of days in billing cycle = 30
Actual use = 57,200 kWh

Hence:

Load factor = {57,200 kWh / (30 × 24 hours per day × 436 kW)} × 100%

= 57200/313920 x 100 =18.22%

The above simply illustrates that at 100% efficiency the usage would have been 313,920 kWh over 30 days but in reality was only 57,200 kWh.

Another and perhaps more pertaining example would be a wind generator, which is not, by its very nature, a very effective provider of continuous electrical energy. The Load Factor of a wind generator in the UK is typically of the order of 29 per cent or less, and although a modern wind generator produces electricity 70-85 per cent of the time, it generates different outputs depending on the wind speed - thus it should be clearly understood that for most of the time the electricity produced is not 'useful electricity' - over the course of a year, the wind generator will typically generate useful electricity for about 29 per cent of the theoretical maximum output.

As a comparison to other means of electrical generation it is informative to carry out the calculations for the load factor of a roof-mounted solar array:

Over a period of 7 years the average solar generation from a 4 kW solar panel system, mounted on the roof of a house in Ceredigion produced 4000 kWh per year. This generation is factual and not a hypothetical or computer-derived set of figures. The data was collected daily on site and recorded on a spreadsheet; thus daily, weekly, monthly and yearly figures were recorded.

If the array generated full capacity for one year then:

$4 \times 10^3 \times 24 \times 365 = 35040 \times 10^3 = 35$ MWh

Actual generation for year = 4,000 kWh (4 MWh)

Thus the load factor for a roof-mounted solar array: 4/35 x 100 = **11%**

Please note this figure demonstrates that solar power is far less efficient than electricity generated from the abomination of wind farms, which begs the question, 'Why is the UK Government so intent on ruining acres and acres of beautiful countryside with such inefficient and unreliable means of electrical generation.

The roof-fitted solar generation was found to be more than acceptable, especially when considering the solar panels were not in the optimum position for maximum generation. It is recognised that west Wales is not the sunniest place, and placement in sunnier England would produce more energy.

Apart from the quality and build of a solar panel, the angle at which the panels are pitched is important. The ideal angle in the UK is about 30 degrees off the horizontal, simply because this allows the panels to extract more from the sun in the morning and the evening, while still making the most of the mid-day sun. For optimum performance, the panels should be facing south, that is, towards the equator in the northern hemisphere. West works well too, and a mix is acceptable – noting the Ceredigion panels were orientated to the south west. A solar panel placed flat onto a west- facing wall will produce about half the amount of electricity compared to being placed at a 30 degree angle on a suitably pitched south facing roof. A north facing roof should, obviously, never be considered.

A total of 16 solar panels were mounted on the roof of the detached property at Ceredigion, which has a tilt of approximately 40 degrees and a south-westerly orientation. The solar array consists of sixteen Hyundai 250 Watt Black Frame Polycrystalline panels offering a total system capacity of 4 kW at a cost of £7,100. The generation from the roof-mounted panel's poses a conundrum, as to why Government has not *fully* embraced this potential, limiting itself to the recently ended Feed-in Tariff (FIT) scheme, which has been replaced by a less attractive Smart Export Guarantee (SEG) Scheme. Pitched or flat roof-fitted solar panels are GREEN energy which avoids the covering and industrialising acres and acres of countryside, with an unsightly multitude of solar panels - being mounted on the top of buildings are less of an eyesore, and many situations can be sympathetically placed, or indeed out of sight as in the case of flat roofs. So why has the Government not considered this scenario and introduced a scheme whereby installation can be offered free, or at a much-reduced price to the public? The basic logic of such a large scheme would be the reduction of less environmentally-friendly means of generation, while at the same time ensuring security of supply. Obviously solar generation does not function at night, or on cloudy days. In the winter, with its short days and overcast skies, the National Grid must have the capacity to keep the lights on. The scheme would need careful integration into the National Grid to achieve a correct balance of different types of generation to avoid blackouts or brownouts.

In relation to the level of expected solar radiation, the following calculations are considered accurate for England and Wales only. The calculations below will show that for the cost of building a large nuclear power station, then 10 million 4 kW solar panel systems could be installed:

A 4 kW roof-mounted solar panel system has clearly shown that it can generate 4,000 kWh per year in an environment not known for its annual sunshine hours.

1 million 4 kW systems each generating 4000 kWh per year will generate:

$$4000 \times 10^3 \times 10^6 = 4{,}000 \text{ GWh per year (4 TWh per year).}$$

Thus 10 million systems will generate:

$$40{,}000 \text{ GWh per year (40 TWh per year).}$$

As an academic exercise it is interesting to note that total UK electricity demand was 346 TWh in 2019, therefore if 20 million dwellings were fitted with 4 kW solar systems then 80,000 GWh (80 TWh per year) would be generated which would result in:

$$80 \times 100/346 = 24.85\%$$

Thus 25% of UK total electricity demand.

At the time of writing a 4 kW solar panel system can be installed for about £5,000, and at those prices for 1 million dwellings it would cost:

$$5000 \times 10^6 = 5 \times 10^9 = \pounds 5 \text{ billion}$$

According to the Office for National Statistics (ONS) at www.ons.gov.uk there were approximately 25 million households in the UK during 2017. If 10 million were considered suitable for solar panels, then the cost for installation for 10 million households would be £50 billion. In tendering for 10 million solar panel systems astute business leaders would be in a great position to easily strike a 50 per cent reduction or less in purchasing and fitting the panels, and therefore be in a very strong position to fit 10 million systems for £25 billion or less. Hopefully there would be some cerebral thinking from Government to recognise and support the potential. As a comparison the cost of Hinkley Point Nuclear Power Station is estimated at £23 billion, at the time of writing. The capacity of the station will be 3.2 GW.

The total capacity of 10 million 4 kW solar panel systems:

$$4 \times 10^3 \times 10 \times 10^6 = 40 \text{ GW}$$

However, Hinkley Point Nuclear Power Station has the ability to generate power 24 hours a day, that is, 8,760 hours per year. The problem is that solar panels are totally dependent upon the Sun, and obviously are powerless at night and usually inefficient on cloudy spring/summer days, especially during the winter months.

Hinkley Point 3.2 GW Power Station has the annual ability to generate:

$$8760 \times 3.2 \times 10^9 = 28,032 \text{ GWh per year (28.032 TWh per year)}$$

Hinkley Point, similar to other power stations, will not generate at full capacity every hour of the day, owing to maintenance and National Grid requirement. Nevertheless it should be noted that nuclear reactors are very reliable at generating electricity, capable of running for 24 hours a day for many months, if not years, without interruption, regardless of the weather or season. Nuclear power stations and coal-fired power stations usually produce the minimum level of electricity required by the National Grid over a period of 24 hours - this is called BASE LOAD electricity. Nuclear and coal-fired power stations are programmed to run all the time, except for maintenance, and National Grid dictates, simply because they take the longest time to start up. It is recognised that although 10 million 4 kW solar panel systems are capable of a total of 40 GWh per year. However, the power may not always be required at a time when a specific customer needs it. This surplus power at any one particular instant will be fed into the National Grid for the use of other customers, and obviously lessen the demand from other generation sources. But of course, if it is not needed at that specific time as a result of energy abundance, then it will be dumped.

In comparing the capital cost and horrendous cost of decommissioning a power station such as Hinkley Point, nuclear energy does not make sense. There is also the ongoing problem of nuclear waste, and while recognising that nuclear power stations are built by the sea for water cooling there are potential problems. Such as the short-sightedness of building a nuclear power station on low lying coast, that was inundated with flood water in 1606. Who is to say a repeat flooding will not occur, but if it does it has the potential to create a nuclear disaster similar to Fukushima, Japan. Sea water was used to cool the Fukushima nuclear plant which went into meltdown in 2011, when a tsunami led to a catastrophic nuclear disaster on Japan's Fukushima coast - but then it may happen only every 400 years or so, but why take the chance when there are other options to nuclear plants as explained in this book? As a bonus this policy will call a halt to nuclear generation and the possibility of a nuclear disaster on our small island. If politicians of all parties considered all the facts, roof-fitted solar energy supplies make a whole lot of sense coupled with enhancing voter confidence. Politicians should be asking Parliament, "Why are we not spending the money more wisely on roof-mounted solar panels, which would result in many happy customers having their power bills reduced, and stop profits going to foreign suppliers of energy." The UKs largest energy suppliers are, EDF Energy, SSE, Npower, ScottishPower and E.ON.

EDF Energy: Formed after the French energy company *Electricite de France* purchased London Energy. It is also one of the largest distribution network operators in the UK after taking control of the UK nuclear generator, British Energy.

SSE: Formed in 1998 after a merger of Scottish Hydro Electric and Southern Electric. SSE's headquarters are located in Perth, Scotland. SSE also includes the sub-brands SSE Atlantic, SSE Southern Electric, SSE SWALEC and SSE Scottish Hydro.

Npower: Bought by the German provider RWE in 2002, and is owned by Innogy SE, a subsidiary of the German company RWE, which is an incorporation of Germany's leading energy companies.

ScottishPower: Formed in 1990 after the state-owned Scottish electricity industry became privatised. Its headquarters may be located in Glasgow but, since 2006, the supplier has been a subsidiary of Spanish utility company Iberdrola.

E.ON: Founded in 1989, known back then as Powergen, this supplier was purchased by German energy company E.ON in 2002. Its headquarters are located in Dusseldorf.

Numerous companies such as Cadbury which is owned by Mondelēz International, a U.S. multinational confectionery, food and beverage conglomerate; Jaguar Land Rover and Asda are among the big UK names owned by foreign companies - even the national lottery are now controlled by businesses abroad. According to the Office for National Statistics (ONS), more than 41 per cent of UK companies are foreign owned.

Air Source Heat Pump

Fundamentally an air source heat pump is a system which transfers heat from outside to inside a building, or vice versa. For domestic heating purposes an air source heat pump absorbs heat from external air and releases the heat energy inside the building, as either hot air, hot water- filled radiators, underfloor heating or domestic hot water supply; the same system can often do the reverse in summer, cooling the inside of the house.

The heating performance and efficiency of an air source heat pump system is commonly measured by the Coefficient of Performance (CoP). The CoP is a simple calculation which works out how much energy the heat pump is able to extract from the energy source compared to the amount of electrical energy it uses. For example:

$$CoP = \frac{\text{Heat output of system (useful heat)}}{\text{Electrical input from compressor and fan motors}}$$

$$= \text{6 kW heat pump/1.2 kW of electrical input}$$

$$= \text{CoP of 5}$$

Generally speaking, the higher the CoP figure, the greater the efficiency of the heat pump. A CoP however only applies to a specific temperature, which means that the CoP rating is not representative of the performance that could be achieved across a whole year. A far more accurate assessment of efficiency therefore is provided by the Seasonal Coefficient of Performance (SCOP). It defines the performance of the heat pump over the course of the year, with seasonal variations in conditions.

Appendix Four

Electric Vehicles (EVs)

Power assessment for 40 million vehicles

Due to the number of unknowns and variables the approach has been to determine the electrical power required to charge 40 million EV batteries from flat, to a full charge - then modify the results appropriately in an attempt to attain realistic charging predictions.

Of course, fully charging 40 million vehicles simultaneously will NOT happen in reality, any more than 40 million petrol/diesel vehicles rushing to filling stations simultaneously to fill their empty tanks.

For simplicity sake it is assumed the 40 million EVs have the Nissan Leaf specification with a battery capacity of 40 kWh. Other data used, with acknowledgement, was published by Statista Research Department, Oct 26, 2020. Digest of UK Energy Statistics (DUKES), Nissan (www.nissan.co.uk), National Grid (www.gridwatch.co.uk) and Department Of Transport, Vehicle Licensing Statistics: Annual 2019.

It should be noted there were 8,380 petrol filling sites in the UK during 2020 having dropped from 13,107 in 2000. According to the Petrol Retailers Association (www.ukpra.co.uk/en/about/facts-figures) the total road fuel for 2019 amounted to 46.5 billion litres, with retail petrol at 16.2 billion, retail diesel 20.8 billion litres, and commercial diesel 9.5 billion litres. At the end of 2019, there were 38.7 million licensed vehicles in Great Britain, a 1.3 per cent increase compared to the end of 2018. Therefore in this evaluation it is assumed there will be a 1.3 per cent increase for 2020 which gives a figure of 39.2 million. Thus for simplicity sake this figure is rounded to 40 Million vehicles.

Prime Minister Boris Johnson said during November, 2020 that new cars and vans powered wholly by petrol and diesel will not be sold in the UK from 2030. This is a time period of ten years and obviously the conversion to electric cars will not take place overnight, but whether the introduction will follow a straight line or exponential curve remains to be seen – although an exponential curve is the most likely scenario if it is to become a reality. It is the author's opinion that Government targets for electric vehicles are unrealistic and therefore not attainable. Hopefully the following assessment of 40 million vehicles will clearly make the case.

According to the Statista Research Department the average motorist in the United Kingdom (2017/2018) drove 10,000 miles per year. As the author has no evidence to prove otherwise, this figure has been used with acknowledgement.

A Nissan Leaf car (cost £26,845 new at November, 2020) has a 40 kWh capacity battery and can travel 168 miles on a full charge, whereas the Nissan Leaf E+ has a 62 kWh battery capacity and can travel 239 miles on a full charge. Nissan offer two options for charging, namely the 7 kW charger at home, office or on the road, and secondly, the 50 kW Chademo rapid charger, which will charge the Nissan Leaf in 60 minutes or the Leaf E+ in 90 minutes from **20 per cent to 80 per cent.** To fully charge one Nissan Leaf car with a fully discharged battery (capacity of 40 kWh) at 7 kW will need 5.7 hours of charging. (It should be noted though that in practice, after the initial full charge, the battery should not be drained to a fully discharged state). Batteries are seldom fully discharged, and manufacturers often use the 80 per cent **Depth-of-Discharge (DoD)** formula to rate a battery. This means that only 80 per cent of the available energy is delivered and 20 per cent remains in reserve. Nissan recommend charging the battery at 80 per cent to preserve battery life, and a battery is restricted to 80 per cent charge only if charged through a rapid charging station (50 kW) to preserve battery life. Numerous factors determine how far an EV will travel on a full charge. Driving on a wet winters evening with lights, wipers and heater on will certainly lessen the range of the car.

As mentioned earlier that for simplicity sake it is assumed that every car has the Nissan Leaf specification.

In calculating the electricity for 40 million EVs there are two important factors to consider.

1. The power (GW) required at any one moment in time (peak).

2. The total electrical energy (TWh) required per annum.

Power demand for 40 million cars charging simultaneously:

$40 \times 10^6 \times 7 \times 10^3 = 280 \times 10^9 = 280$ GW (280,000 MW).

Peak demand for electricity in the UK for 2019 was 47.275 GW, which will mean that if 40 million EVs charge their batteries simultaneously, and coincide with UK peak demand, then the power demand will be:

$47.275 + 280 = 327.275$ GW

Resulting in 7 times the 2019 peak demand of the National Grid:

327.275/47.275 = 6.92

But of course, in reality, 40 million EVs will not be charging all at the same time, nor will they be charging a flat battery to a full battery, but charging between 20 per cent and 80 per cent of battery capacity. Additionally, all 40 million EVs will hardly be charging their batteries during normal Grid peak demand.

Ofgem has called for incentives to encourage people to charge their EVs outside of peak hours, which they say would increase the number of cars currently supportable by the country's electricity network by 60 per cent. This is very similar to the National Grid's own opinion, which is that flexible charging would halve the estimated additional generation needed to manage the demand. National Grid on their website (www.nationalgrid.com) say there could be 36 million electric vehicles on the roads by 2040 and through *smart charging technologies*, charging vehicles at off-peak times and through vehicle-to-grid technology, the increase in peak demand from vehicles could be as little as 8 GW.

The author certainly challenges the figure of 8 GW (unless consumers are being purposefully cut-off), and deems the figure extremely low due to the following: Again for simplicity sake the Nissan Leaf has been chosen as the 'standard vehicle' for the average annual mileage of the UK motorist at 10,000 miles. Thus if the motorist covers 10,000 miles per year, and a Nissan Leaf car travels 168 miles before depleting its battery, then to satisfy the 10,000 miles per annum, each driver it will require a total of:

10,000/ 168 = 59.5 charges.

To fully charge a completely flat 40 kWh capacity battery, using a 7 kW charger will take:

40/7 = 5.7 hours to fully charge.

If the battery is depleted by 80% it will require a total charge of 32 kWh, using a 7 kW charger and will take:

40 x 0.80/7 = 32/7 = 4.57 hours to fully charge.

But if depleted by 20% it will take:

40 x 0.20/7 = 8/7 = 1.1 hours to fully charge.

Therefore depending on the state of the battery the charging times can vary between 1.1 to 4.7 hours, and the demand on the Grid will vary directly as a consequence of different charging times. But in reality, apart from the Nissan Leaf, and different charger outputs ranging from 3 kW to the 50 kW Chademo rapid charger, the situation will be much more complicated due to the range of various electric vehicles and other charging rates that will be available. It is deemed not unreasonable to suggest the average driver will travel no further than a total of 25 miles for a journey trip to work and back. Thus the Nissan Leaf car can make this journey: 168/25 =

6.72 times - possibly with many drivers making 6 journeys before charging the vehicle battery. Of the 40 million vehicles it is expected that many will travel short distances, whilst others longer trips. Considering the many unknowns, and to attempt a meaningful and conservative prediction, it is assumed that 25 per cent of the 40 million cars in 2030 will require charging on arriving home, and they will need a 50 per cent charge, that is, a 20 kWh charge at 7 kW for 2.85 hours.

If 10 million cars simultaneously charge their batteries at 7 kW then:

$$10 \times 10^6 \times 7 \times 10^3 = 70 \text{ GW}$$

This demand will last for about 3 hours and will mean an additional 70 GW (70,000 MW) of power from the Grid.

To put this into context this is equivalent to the maximum power output of 35 conventional large power stations rated at 2 GW. Pembrokeshire CCGT power station has a capacity of 2 GW. Electricity demand is low during the night, with little domestic or commercial consumption - there is a surge in demand in the morning, when people wake-up and start to use items such as kettles, toasters, microwaves and electrical showers – electricity demand continues to rise but then starts to stabilise at around 0900 hrs as offices and shops open and electrical equipment such as computers are switched on. In the winter a second surge then occurs later in the day, between 1530 hrs and 1730 hrs, as school children begin to return home and the working day starts to come to an end. When adults return home they will be turning on electrical items such as lights, televisions, kettles, ovens et cetera and begin to start cooking the evening meal. Electricity demand then falls as people begin to retire to bed. Unfortunately if 10 million EV charging occurs at the same period as the Grid peak time of 1800 hours to 2000 hours, when most people are cooking dinner, switching on lights, watching TV, and using various appliances around the home. This will result in a **new peak demand** of:

70 GW + 48 GW = **118 GW**

(Note: the 2019 peak demand of 47.275 GW has been rounded to 48 GW.)
If it is deemed that 25% is a high figure, then a figure of 10% (4 million EVs) charging simultaneously will still result in 28 GW:

$$4 \times 10^6 \times 7 \times 10^3 = 28 \text{ GW} \quad \text{(With a peak of 28 GW + 48 GW = 76 GW)}$$

This extra power is equivalent to FOURTEEN large power stations generating 2 GW and working at full capacity. In reality 2 GW gas-fired power stations have a Load Factor (see appendix three and glossary for definition) in the order of 60 per cent which would result in at least TWENTY TWO such power stations.

The National Grid figure of **8 GW** for 36 million EVs is thus challenged as how is this possible when 4 million EVs will require **28 GW**, and surely for 36 million EVs the requirement will be **252 GW**:

$$36 \times 10^6 \times 7 \times 10^3 = \textbf{252 GW}$$

The calculations indicate the National Grid figure is totally unrealistic, unless new and untested ways are developed of spreading the power required across the whole day. However the difference between **8 GW** and **252 GW** is so stark that such developments appear highly unlikely.

Annual UK electricity

For a prediction of the total electricity consumed per annum required by 40 million EVs, it will be helpful to know how far each vehicle will travel in a period of one year. The following calculations are based on the average UK motorist driving 10,000 miles per year.

The Nissan Leaf car can travel 168 miles on a full charge and therefore 40 million EVs will require:

$10000/168 = 59.5$ full charges.

The Nissan Leaf fully charged battery has a capacity of 40 kWh, thus 59.5 full charges to satisfy 40 kWh for each battery will require:

$59.5 \times 40 \times 10^3 = 2,380$ kWh per annum

This is approximately half the current average annual domestic power bill, which means the average annual bill, will increase by 50 per cent (when vehicles charged at the normal domestic rate).

40 million fully charged 40 kWh batteries will require:

$$40 \times 10^6 \times 2380 \times 10^3 = 95.2 \text{ TWh}$$

Therefore the additional annual electricity requirement = 95.2 TWh

The UK consumed 346 TWh during 2019, thus for 40 million cars the total amount of total power required will be 95.2 + 346 = **441.2 TWh.**

To put this extra energy (95.2 TWh) into context it should be recognised that for a period of one year a 2000 MW power station working continuously (100 per cent Load Factor) has the capacity to generate:

$$2000 \times 10^6 \times 8760 = 17.520 \text{ TWh}$$
Note: hours in one year = 365 x 24 = 8760 hours

Thus 95.2 TWh would require *at least* an additional SIX power stations:

$$95.2/17.52 = 5.43$$

But of course, in reality, power stations such as a 2 GW gas-fired power station have a load factor of about 60% therefore:

$$95.2 \times 10^{12}/2 \times 10^9 \times 8760 \times 0.60 = 95.2/17.52 \times 0.60 = 9$$

Thus the additional power stations will be at least **NINE.**

Summary

In relation to peak demand the calculations have shown, theoretically, that should 40 million EVs charge simultaneously and coincide with a Grid peak of 47.237 GW then a total 327 GW will be required. Of course this is a long way from *reality* as 40 million vehicles will not all be charging at the same time, and it is somewhat problematic, due to the unknowns and variables, to predict a peak charging time for EVs. Nevertheless, a '*best guess*' figure suggests the peak would be in the range of 28 GW to 70 GW. Thus the author feels fairly confident in suggesting peak demand for 40 million EVs will necessitate a possible **50 per cent increase** (23.64 GW) of the UK current peak demand of 47.275 GW. The National Grid 8 GW is regarded as a tongue-in-cheek figure. If the lower '*best guess*' figure of 28 GW transpires to be the most accurate then that would require an additional **FOURTEEN** 2 GW power stations.

Regarding annual electricity consumption for 40 million EVs the author feels more confident in predicting an additional annual need of 95.2 TWh, which will mean a new annual consumption of 441.2 TWh as compared to the current 346 TWh. This extra consumption will translate in the need for at least **NINE** additional 2 GW sized power stations. This would suggest that the minimal number of additional 2 GW size power stations, would be in the region of **NINE to FOURTEEN** 2 GW power stations.

Wind Generation for EVs

The Load Factor of onshore wind generators in the UK is typically of the order of 26 per cent or less, and although a modern wind generator produces electricity 70 per cent to 85 per cent of the time, it generates various outputs at different times depending on the wind speed. Thus it should be clearly understood that for most of the time the electricity produced is not *'useful electricity'* - over the course of a year, the wind generator will typically generate useful electricity for about 26 per cent of the theoretical maximum output. As a comparison and according to 'Statista' the Plant Load Factor (PLF) of combined cycle gas turbine (CCGT) stations in the United Kingdom has fluctuated since 2010. In 2019, PLF of combined cycle gas turbine stations was just over 60 per cent. It should be clearly recognised the output of a fossil-fuelled power station is predominantly controlled by man, and will offer a secure supply, whilst that of a wind generator is dependent on the elements (wind) and therefore cannot offer any security of supply.

The UK consumed 346 TWh of power in 2019, and the peak demand was 47.275 GW (DUKES). The following calculations show how many large 2000 MW CCGT Power Stations, and a 2 MW wind generator working continuously at 100 per cent efficiency would be required to satisfy a peak demand of 47.275 GW:

Peak demand 2000 MW power stations

$47.275 \times 10^9 / 2000 \times 10^6 = \textbf{23.6}$

Thus to satisfy a peak of 47.275 GW would require 24 power stations at full generation.

Peak demand 2 MW Wind generators

$47.275 \times 10^9 / 2 \times 10^6 = 23.63 \times 10^3 = 23,637$

Thus to satisfy a peak of 47.275 GW would require 24,000 wind generators working flat out.

Unfortunately CCGT power stations and wind generators do not work at 100 per cent capacity, but are restricted by their Load Factors. The Load factor of a CCGT power station is of the order of 60 per cent whilst that of onshore wind for 2019 was 26 per cent and offshore 40 per cent.

Power station at 60 per cent load factor:

$47.275 \times 10^9/2000 \times 10^6 \times 0.60 = 47.275/1.2 = 39$ power stations

Onshore wind generator at 26% generation:

$47.275 \times 10^9/2 \times 10^6 \times 0.26 = 47.275 \times 10^3/0.52 = 90,913$ wind generators

Offshore wind generator at 40% generation:

$47.275 \times 10^9/2 \times 10^6 \times 0.40 = 47.275 \times 10^3/0.80 = 59,093$ wind generators

Thus we have:

CCGT power station at 60% = 39 power stations.

Onshore wind generators at 26% = 90,913 wind generators

Offshore wind generators at 40% = 59,093 wind generators

Number of power stations for 346 TWh

2 GW Power station

At 100% Load factor:

$346 \times 10^{12}/2000 \times 10^6 \times 8760$

$= 346/2 \times 8.760 = 19.75$

Hours in a year $= 365 \times 24 = 8760$ hours

This results in 20 power stations.

At 60% Load factor:

$346 \times 10^{12}/2000 \times 10^6 \times 8760 \times 0.60$

$= 346/17.52 \times 0.43 = 346/7.53 = 33$

This results in 33 power stations.

2 MW generators for 346 TWH of power

At 100% Load Factor:

$2 \times 10^6 \times 8760 = 17,520$ MWh (17.520 GWh)

$346 \times 10^{12}/17.520 \times 10^9 = 19.75 \times 10^3 = 20,000$

This results in 20,000 wind generators

At 26% Load Factor:

$346 \times 10^{12}/17.520 \times 10^9 \times 0.26 = 346 \times 10^3/4.55 = 76,000$ wind generators

At 40% Load Factor:

$346 \times 10^{12}/17.520 \times 10^9 \times 0.40 = 346 \times 10^3/7 = 49,000$ wind generators

If the country were to be totally reliant on wind generation then 76,000 onshore or 49,000 offshore wind generators would be required.

Comparing the power stations and wind generators required for peak demand and annual demand we have:

	Load factor	Peak demand (47.275 GW)	Annual demand (346 TWh)
Power Station	60%	39	33
Wind generator	26%	90,913	76,000
Wind generator	40%	59,093	49,000

To put this into perspective Whitelee Wind Farm near Eaglesham, East Renfrewshire, at the time of writing, is the largest ONSHORE wind farm in the United Kingdom with 215 Siemens and Alstom wind generators and a total capacity of 539 MW. Clyde Wind Farm near Abington, South Lanarkshire is the UK's second largest onshore wind farm comprising 152 generators with a total installed capacity of 350 MW.

If Whitelee Wind Farm has a capacity of 539 MW from 215 wind generators means that each generator has a capacity of:

$539/215 = \textbf{2.5 MW}$

Thus a 346 TWh consumption employing 2.5 MW wind generators working at 40 per cent generation would require:

$2.5 \times 10^6 \times 8760 = 21,900$ MWh (21.9 GWh)

$346 \times 10^{12}/21.9 \times 10^9 \times 0.40 = 346 \times 10^3/8.76 = 39,497$ wind generators

$39497/215 = 184$ wind farms the size of Whitelee Wind Farm.

Therefore it will require 184 wind farms each consisting of 215 wind generators each having a capacity of 2.5 MW enabling them to generate 346 TWh per annum. But of course, under a large windless high pressure system over the UK these 184 wind farms will NOT be generating a single kilowatt of electrical power, if the wind is too strong then wind generators have to shut down to avoid structural damage – such is the complete unreliability of wind generation and its innate inability to deliver a secure power supply.

During 2019 the UK peak demand for electricity was 47.275 GW, but Boris Johnson has committed the Government to produce 40 GW from offshore wind farms by 2030, which will be 85 per cent of UK demand at peak. Although offshore wind generators generally have a higher Load Factor, it still translates to the country facing many brownouts and blackouts - unless a guaranteed source of additional power is factored in. Not to overlook the fact that if 40 million EVs actually become a reality, coupled with the nation being forced to use heat pumps, then the country will be in serious trouble.

Assessment using Ofgem and Internet data

The average household according to Ofgem uses 3,100 kWh per year, and researching the Internet it seems the change to EVs will result in each household using an extra 2,000-3,000 kWh per year in electricity. Noting earlier calculations in this appendix revealed 2,380 kWh per year.

The Office for National Statistics (ONS) records show there were approximately 25 million households in the UK during 2017. Thus to supply the lower figure of an extra 2,000 kWh per year the UK will need to generate:

$25 \times 10^6 \times 2000 \times 10^3 = 50 \times 10^{12} = 50$ TWh per year

As we have seen earlier a 2 GW power station at 100 per cent load factor has the capacity to generate:

$2000 \times 10^6 \times 8760 = 17.520$ TWh per annum
Note: hours in one year $= 365 \times 24 = 8760$ hours

But at 60 per cent load factor will generate:

$2000 \times 10^6 \times 8760 \times 0.60 = 2 \times 8.760 \times 0.60 \times 10^{12} = 10.512$ TWh

Thus, $50/10.512 = 4.756$

This will entail building FIVE additional 2 GW power stations:

If the household extra demand is 3,000 kWh per year then:

$25 \times 10^6 \times 3000 \times 10^3 = 75$ TWh per year

Thus, $75/10.512 = 7.13$, results in SEVEN additional 2 GW power stations at 60 per cent load factor.

Note: Earlier calculations relating to EV demand revealed a requirement for SIX such power stations at 100 per cent load **factor** to produce 95.2 TWh per year.

Considering a 2 GW gas-fired power station costs in the region of £1 billion, then between £5 billion and £7 billion will need to be funded, depending on which calculations prove to be correct.

The number of wind generators for 25 million households needing an extra 2,000 kWh per year, will result in an annual demand of:

$25 \times 10^6 \times 2000 \times 10^3 = 50$ TWh per annum

The annual generation of a 2 MW wind generators working at 26 per cent efficiency is:

$2 \times 10^6 \times 8760 \times 0.26 = 4,555$ MWh per year

The number of 2 MW generators at 26 per cent efficiency for 25 million households needing an extra 2,000 kWh per annum:

$50 \times 10^{12}/4555 \times 10^6 = 11,000$ wind generators

The number of 2 MW generators at 26 per cent efficiency for 25 million households needing an extra 3,000 kWh per annum:

$75 \times 10^{12}/4555 \times 10^6 = 17,000$ wind generators

The number of 2 MW wind generators working at 26 per cent efficiency to satisfy total UK annual usage of 346 TWh has been earlier been determined at 76,000, and thus would corroborate the household calculation:

Household requirement 50 TWh = 11,000 generators

Total UK requirement = 346 TWh = 76,000 generators

Thus 346/50 = 6.92, giving a factor of 7 therefore:

11,000 x 7 = 77,000 wind generators.

Therefore the UK wind generation requirement would require AT LEAST 76,000 wind generators each having a 2 MW capacity with a load factor of 26 per cent.

At the time of writing the UKs largest OFFSHORE wind farm, The London Array, had a total of 175 generators, each with a capacity of 3.6 MW, offering a total of 630 MW at 100 per cent load factor. The total number of wind generators in the UK at the time of writing, are 8,600 onshore and 2,300 offshore, giving a total of 10,900 generators. The total capacity of UK wind generation is 24.1 GW (24,100 MW).

The calculations have indicated that it will require at least seven times the current number of wind generators, which will require a mind boggling area of land and sea needed to accommodate these weather dependent and limited bird and bat mincing machines! All to satisfy a badly conceived Government policy of putting the countries energy requirement in *a house built on sand*, which will only end in tears, unless halted.

Infrastructure

A significant hurdle to the realisation of electric vehicles in the UK will be the availability of suitable charging points especially for drivers who work and live in areas where there are high-rise office buildings, tower blocks and tall apartment flats.

Will Government put pressure on places such as supermarket to provide the necessary charging spaces? If so, there will be the inevitable loss of supermarket car parking spaces when compared to petrol and diesel cars. As an example, my local supermarket has provided 4 car parking spaces for EVs, but this has necessitated in taking up 6 car park spaces, using two spaces for the charging equipment. Therefore the available car parking space has been reduced by a third, meaning a large supermarket with 900 car parking spaces could possibly be reduced to 600 car parking spaces.

A further headache will be the capacity of the cables feeding such car parks. If a 600 space car park decides to have a 100% capacity of charging units of 7 kW output then at any one moment, if all chargers were in use, then the cables feeding these charging units will have to handle:
600 x 7 kW = 4,200 kW (4.2 MW).

Planners would also need to consider how many rapid charges at 50 kW, could be installed adding to the cable handling properties. In reality I doubt if supermarkets, without pressure from Government, will offer this 100% facility, but even if they drop the charging option to a quarter there is still some considerable loss of car parking space and a serious upgrade of power feeder cables to be considered.

It is interesting to note that National Grid on their website (www.nationalgrid.com) state that, 'Over 40 per cent of people in England and Wales do not have off-street parking'.

An aerial survey will quickly identify the numerous terraced houses and high rise flats there are in towns and cities. Consider the scenario with charging points for 40 million cars - streets and roads becoming a labyrinth of cables – a tangled and impenetrable suburban spider's web - cables dangling down from high rise flats, and criss-crossing in streets that have terraced houses. Indeed, in the real world how many EVs can be accommodated with street/road charging facilities? If it is envisaged that charging cables will run across Public Rights of Way such as pavements, then they will prove to be extremely hazardous and dangerous, especially for the disabled and elderly. Will the lawyers have a field day making a fortune from illegal trip hazards? What safeguards, if any, could be put in place to avoid such bone shattering accidents? All this after the streets and roads having been dug up to accommodate network buried cables to the street charging points!

The existing local networks were not designed and built for the extra loading that EVs will inevitably bring, even 20 million EVs will place serious demand on the network and unless the necessary upgrades are implemented the consequences will be overheating cables, switchgear and transformers, resulting in plant failure followed by brownouts and blackouts. Could this network *'Sword of Damocles'* possibly be one of the reasons why Government and the power companies are pursuing the fitting of Smart Meters, as they will offer the potential of disconnecting consumers from a remote computer when parts of the network are threatening over-demand for electricity – EV owners will not be happy bunnies if there is no power available to charge their vehicles when required. It also disingenuous to claim that Smart Meters are being fitted to save the customer electricity when the power companies are in the market to SELL their product, which of course is electricity!

Appendix Five

Electricity from space

Waves: Electromagnetic radiation (EM radiation) travels in waves at the speed of light. Unlike waves that travel through sound and water, EM waves require no medium. They can move through air as well as the vacuum of space.

Frequency: The frequency of a wave is measured in Hertz (Hz). 1 hertz is equal to one cycle per second of the wave.

The electromagnetic spectrum: This refers to the range of all types of EM radiation, which is a form of energy. The difference between one end of the spectrum and the other is determined by the frequency of the waves. Visible light makes up one section of the EM spectrum, as do radio, X- rays and gamma rays.

Frequency bands: This term simply refers to the chunks of wavelengths making up the spectrum. Ka-band, often used for satellite, is one type of band. Visible light is another. Some bands are quite large, while others may have just a *sliver* of 'bandwidth'. AM radio operates at a frequency between 535-1605 kilohertz (kHz), so a station at 800 kHz has waves cycling 800,000 times per second. A signal from a Ka-band satellite operates at a much higher frequency of around 28 gigahertz (GHz), 28,000,000,000 times per second.

GEOSYNCHRONOUS satellites

Used for communicating back and forth with spacecraft (such as the Hubble Space Telescope and space shuttles), voice communication. It is a satellite that remains in geosynchronous orbit around our planet - its orbital period is the same as the Earth. In other words, a geosynchronous satellite revolves around the planet at the same speed at which the planet rotates on its axis. That is the reason why this type of satellite appears to be in the same region in the sky (at a given time of the day) when viewed from a particular position on Earth.

The orbital period of a geosynchronous satellite is a sidereal day (23 hours, 56 minutes and 4 seconds), which is why it appears to stay in place over a single longitude (but it may drift south/north depending upon the orbit's inclination with Earth's equatorial plane). Geosynchronous orbits that are circular in shape have a radius of 26,199 miles (42,164 km).

GEOSTATIONARY satellites

Used for global communications and weather forecasting. Examples of geostationary satellites are the American GOES (Geostationary Operational Environmental Satellite) series, the Indian INSAT satellites, Japanese Himawari, and European Meteosat. Many folk get confused between geosynchronous and geostationary satellites - a geostationary orbit (also known as a geostationary Earth orbit, geosynchronous equatorial orbit, or simply GEO) is a circular orbit located at an altitude of 35,786 kilometres (22,236 miles) above the surface of Earth with zero inclination to the equatorial plane. A satellite in this orbit has an orbital period of one sidereal day (23 hours, 56 minutes and 4 seconds), which means that it completes one revolution around Earth in exactly the same time as Earth completes one rotation on its axis. Since a geostationary satellite has the same orbital period as Earth, and it also travels from west to east (the direction in which Earth rotates on its axis), it therefore appears to hover at a single point in the sky when observed from a given point on the ground. Hence, the name 'geostationary', as it appears 'stationary' from a given geographical location.

Comparing geostationary and geosynchronous orbits, it should be noted that there is very little difference between the two. A satellite in geosynchronous orbit has the same orbital period, i.e., one sidereal day, as that of a satellite in a geostationary orbit - the only difference between the two is that while a geosynchronous satellite may or may not be following an inclined orbit (with respect to the equatorial plane) a geostationary satellite has to follow a non-inclined orbit - meaning a geostationary satellite remains exactly above the Earth's equator at all times.

Examples of geosynchronous satellites (which have the inclined orbit) include the Russian Raduga 29, Indian Astra 1C, Malaysian MEASAT 2 and many others.

RECTENNA

A rectenna is a rectifying antenna, which is a special type of receiving antenna that is used for converting electromagnetic energy into direct current (DC). They are deployed in wireless power transmission systems that transmit power by radio waves. A simple rectenna element consists of a dipole antenna with an RF diode connected across the dipole elements. The diode rectifies the alternating current (AC) induced in the antenna by the microwaves, to produce DC power, and powers a load connected across the diode. Schottky diodes are used as they have the lowest voltage drop and highest speed and therefore have the lowest power losses due to conduction and switching. A large rectenna consists of an array of a number of such dipole elements.

Glossary of Terms

Selection of useful terms

AC: Alternating current.

Alternating Current: An electrical current that reverses direction at a regular rate. In the United Kingdom this happens at fifty times a second and is known the frequency; see also Frequency and Hertz.

Aerial: A conductor or arrangement of conductors radiating or collecting electromagnetic energy.

Ammeter: An instrument for measuring the amount of current in an electrical circuit.

Amperage: The quantity of electrical current flowing in an electrical circuit.

Ampere: Unit of current. A current of 1 ampere represents the movement of 6,280,000,000,000,000,000 electrons past a given point in a circuit during 1 second of time.

Armature: The moving part or parts of an electrical motor or generator - also the moving part of a relay, bell or buzzer.

Atom: The smallest particle of an element that has all the element's chemical properties, composed of a nucleus and a number of surrounding electrons.

Atomic nucleus: The central part of an atom consisting of protons and neutrons. The protons have a positive charge, giving the nucleus a positive electric charge; the neutrons have no electric charge.

AWEA: American Wind Energy Association.

Base load: This is the minimum amount of electrical demand needed over a 24-hour time period; it is also known as continuous load.

Bond: Usually refers to a conducting bond by which the lead sheath and the armour of one or more cables, or the casing or framework of electrical apparatus/machinery are electrically connected together and/or to earth.

Battery: A group of electrical cells connected in series or parallel.

BERR: Department of Business, Enterprise and Regulatory Reform, formerly DTI.

Blackout: A blackout is a total crash of the power grid due to an imbalance between power generation and power consumption.

BNFL: British Nuclear Fuels.

Brownout: The term originates from the dimming of incandescent lighting when experiencing a voltage reduction. A brownout can be a selective power cut and controlled shutdown of the power supply in a given area, so as to avoid a blackout. Intentional brownouts are used for load reduction in an emergency. A voltage reduction may be an effect of an unexpected disruption of an electrical grid, and this is known as an unintentional brownout.

Brushes: Devices that provide stationary connections to the rotor in an electrical generator or motor. Carbon is commonly used for brushes in electrical hand tools.

Bulb: See lamp.

BWEA: British Wind Energy Association.

Cable: An arrangement of stranded conductors to form one common core/conductor and insulated throughout its length. A number of cores may be enclosed in a protective sheathing – which may be further protected by armour.

Capacity factor: American terminology for load factor, see also load factor.

CASSIOPeiA: Abbreviation for Constant Aperture, Solid-State, Integrated, Orbital Phased Array.

Cathode: The negative terminal of a cell or battery, also the plate/source in a thermionic device from which electrons are emitted.

CCGT: Abbreviation for Combined Cycle Gas Turbine - power station that uses natural gas.

Cell: A source of electrical energy dependent upon chemical action. A voltaic cell is made of two different kinds of conductor materials placed within a paste or fluid (called an electrolyte) that also conducts electricity.

Coil: Turns of wire conductor to concentrate a magnetic field.

Circuit: An arrangement of conductors and components connected together to carry an electric current.

Circuit Breaker: A switch for making and breaking an electrical circuit under normal or fault conditions.

CME: Abbreviation for coronal mass ejection.

Co-generation: The generation of electrical energy and usable heat in the form of hot water or steam, from the same quantity of fuel in a single operation.

Conductor: A substance which offers low resistance to the flow of an electrical current: a solid, liquid or gas, through which electrons can pass easily - gold, silver and copper are good electrical conductors.

Converter: A device for converting power from AC to DC and vice versa.

Current: The flow of electricity around an electrical circuit.

DC: Direct current.

DECC: Department of Energy and Climate Change.

Direct Current: An electrical current that flows always in the same direction.

Distribution network: The system (low voltage) which consumer services are connected.

DTI: The old Department of Trade and Industry.

Dukes: Digest of UK energy statistics.

Dumping power: Surplus electric power in excess of existing local load requirements.

Earthed Circuit: An electrical circuit, in which one or more points are connected to earth, see also Ground.

Efficiency: The ratio of energy output to energy input – usually expressed as a percentage.

EDF: Electricité de France (EDF Energy).

Electricity, Dynamic: A form of energy present when electrons move through a circuit.

Electricity, Static: A form of energy present within the space between two oppositely charged objects.

Electricity Meter: An instrument in the consumer's premises which totals up the electrical energy supplied over a given time.

Electrode: A conducting body employed to pass an electric current into and out of an electrolyte, gas or electronic valve/tube.

Electromagnet: A magnet produced by an electric current moving through a coil wound around a core of iron.

Electromagnetism: The magnetism produced by an electric current.

Electromagnetic Induction: The production of an electromotive force by changing magnetic fields.

Electron: A negatively charged particle in the shell of an atom; electrons spin about the nucleus of an atom.

Electrostatic Field: The space around a charged body where the lines of electric force may be detected.

E.M.F: Short for electromotive force. It is the electrical force that causes electrons to move through an electrical conductor/circuit and is commonly known as voltage.

E.on UK: Subsidiary of E.ON Nnetz. (Formerly Powergen).

E.ON Nnetz: German based power company.

EWEA: European Wind Energy Association.

Frequency: The number of cycles completed over a period of time. Used to be expressed in cycles per second but now known as Hertz – one Hertz equals one cycle per second.

Fuse: A protective device to give protection to a circuit or circuits against excessive currents.

Generator: A machine that converts mechanical energy to electrical energy.

Gigawatt: One thousand Megawatts usually abbreviated to GW.

Grid: The network of high-voltage transmission lines.

Ground: Also known as Earth – an electrical connection/path between an electrical circuit and the earth.

Hertz: The unit of frequency abbreviated as Hz. One Hertz is equal to one cycle per second.

Hydroelectricity: The means by which electrical energy is generated by water turbines.

IC: Short for Integrated Circuit. Also called microchip, or chip. An assembly of electronic components, fabricated as a single unit, in which miniaturized active devices (transistors and diodes) and passive devices (capacitors and resistors) and their interconnections are built up on a thin substrate of semiconductor.

Insulator: A material which offers a high resistance to an electric current.

Inverter: An electronic device or circuit, which converts DC to AC. The input DC voltage is usually lower, while the output AC is equal to the Grid supply voltage of either 120 volts, or 240 Volts depending on the country. Inverters are employed as standalone equipment for applications such as solar panel systems.

ISS: International Space station.

Kilovolt: One thousand volts, usually abbreviated to kV.

Kilowatt: Unit of power. One thousand watts usually abbreviated to kW.

Kilowatt-hour: Unit of electrical energy usually abbreviated to kWh.

Lamp: A device for converting electrical energy to light energy.

LCD: Liquid-crystal display.

LED: Light-emitting diode.

Load: The electrical power carried by a circuit or taken from a generator; the amount of electricity used by a device when connected to an electrical supply.

Load Factor: The actual amount of electricity produced by a generator/s compared to the maximum amount possible over the same period of time, and expressed as a percentage: known as Capacity Factor in the USA.

Megawatt: One thousand kilowatts, usually abbreviated to MW.

Nuclear Power: The means by which electrical energy is produced by nuclear reaction; heat from a nuclear reactor is used to produce steam to drive a steam turbine.

Ohm: The unit of electrical resistance.

Power: The amount of electrical energy delivered in a unit of time.

Power Station: One or more large electrical generators housed in a large building.

Primary Cell: A chemical cell that cannot be recharged.

PV: Photovoltaic is the field of technology and research related to the application of solar cells for energy by converting sunlight directly into electricity.

Rectifier: A device that allows current to flow in one direction only and is used to convert alternating current into direct current.

Resistance: The opposition to current flow through a conductor or circuit; the unit of resistance is the ohm.

Resistivity: The electrical resistance of a conductor of unit cross-sectional area and unit length, and is useful in comparing various materials on the basis of their ability to conduct electric currents.

Rotor: The rotating part of an AC machine.

Series Connection: Conductors, Resistances or circuits are said to be in series when they are connected so that the same current flows in each conductor, resistance or circuit.

SPS: Space based Solar Power Satellites.

Spinning Reserve: Additional generating capacity, available by increasing power output from generators that are already connected to the system.

Sub-station: A point in an electrical supply area at which electricity is supplied in bulk. The electricity is then, via switchgear, transformers and cables directed to suit the system of supply to the particular area. It is important to note that sub-stations do not generate electricity.

Superhet: Superheterodyne radio receiver.

Switch: A device in an electrical circuit that allows (closed) or stops (open) the flow of an electrical current.

Terawatt: One thousand Gigawatt usually abbreviated to TW.

Transformer: A device employing electromagnetic induction to transform alternating power in one winding (primary winding) to alternating power in another winding (secondary winding) usually at different values of current and voltage.

Transmission: The supply of electrical energy at high voltages, usually 132 kV and higher.

Transmission Line: Power lines carried on tall towers called pylons to supply electrical energy at high voltages, see also transmission.

Transponder: Is a radio communications, monitoring, or control device that picks up and automatically responds to an incoming signal. The term is a contraction of the words transmitter and responder. Transponders can be either passive or active.

TRF: Tuned Radio Frequency.

Waves: Electromagnetic radiation travels in waves at the speed of light, unlike waves that travel through sound and water. Electromagnetic waves require no medium - they can move through air as well as the vacuum of space.

WPT: Wireless power transfer.

WET: Wireless energy transmission.

Volt: Unit of electromotive force or potential difference.

Voltage: The electromotive force between different points in an electrical circuit measured in volts.

Voltmeter: An instrument used to measure voltage.

Watt: Unit of power usually abbreviated to W.

Sources

Efergy Elite wireless electricity monitors, (www.efergy.com).

G.B. National Grid Status Website, (www.gridwatch.templar.co.uk), or (www.gridwatch.co.uk).

National Grid, (www.nationalgrid.com).

Marlec Engineering Co Ltd, (www.marlec.co.uk).

BERR. Department of Business, Enterprise and Regulatory Reform, formerly DTI.

GOV.UK (www.gov.uk/guidance/smart-meters-how-they-work). (www.gov.uk/guidance/wave-and-tidal-energy-part-of-the-uks-energy-mix)

Citizens Advice. (www.citizensadvice.org.uk/consumer/energy).

Environmental Protection UK, 1 Samian Way, Stoke Gifford, Bristol, BS34 8UQ. (www.environmental-protection.org.uk/).

Department of Energy and Climate Change (DECC).

Digest of UK Energy Statistics (Dukes).

BEIS, Department for Business, Energy and Industrial Strategy.

The Office for National Statistics (ONS).

CCL Components Ltd, 1 Cairn Court, East Kilbride, Glasgow.

Powerwatch. (www.powerwatch.org.uk).

Solar iBoost, Marlec Engineering Co Ltd, Rutland House, Trevithick Road, Corby, Northants, NN17 5XY. (www.marlec.co.uk).

Petrol Retailers Association. (www.ukpra.co.uk/en/about/facts-figures).

International Union for the Conservation of Nature (IUCN). (www.iucn.org).

Dr Adam Rutherford, geneticist at University College, London,

Orbital Marine Power, Hatston Pier Road, Crowness Business Park, Kirkwall, KW15 1ZL.

BBC News.

The BBC News on-line, (www.bbc.co.uk).

Mark Duchamp, President, Save the Eagles International, (www.savetheeaglesinternational.org).

Dŵr Cymru (Welsh Water), Linea, Fortran Road, St. Mellons, Cardiff, CF3 0LT. (www.dwrcymru.com).

Wikipedia, the free encyclopedia.

World Energy Forum

Organisation for Economic Co-operation and Development (OECD)

Veissmann Limited, Hortonwood 30, Telford, TF1 7YP. (www.viessmann.co.uk).

National Energy Action, The national fuel poverty charity (www.nea.org.uk)

Which? www.which.co.uk/reviews/feed-in-tariffs/article/feed-in-tariffs/what-was-the-feed-in-tariff-aAsa36S95iJy

OTHER BOOKS BY THE AUTHOR

21st CENTURY ELECTRICITY
(Foreword by David Bellamy OBE)
ISBN 978-1-78507-390-8

Obtainable from Amazon or New Generation Publishing

COUNTDOWN TO OBLIVION

Obtainable from Amazon or

Trafford Publishing
ISBN 141202685-7
www.trafford.com
E-mail sales: sales@trafford.com

KINDLE

Cartoon books

I just love my computer

Computer Rage

Romans, Greeks, Egyptians and Gauls

Aliens and Space

Children's fantasy book

Troll Castle and the Forbidden Chamber of Gold

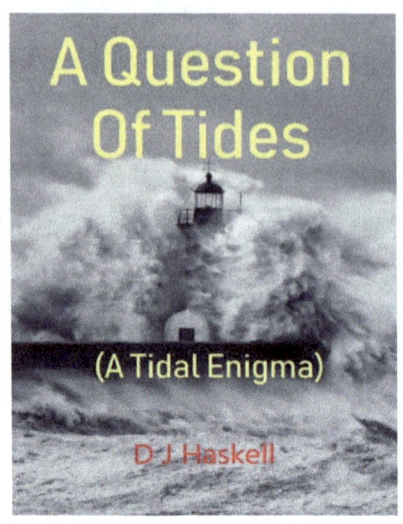

A Question of Tides

ISBN 978-0-244-76924-6

Obtainable from Lulu Publishing (www.lulu.com)
Priced £5.99

Amazon priced £5.99 Kindle Edition (EBook) £2.93

It is enigmatic that many people do not appreciate the ocean tides are due to the Earth falling toward the Moon, in the same manner as the International Space Station falls around the Earth. All is revealed in an easy to understand manner. The book also clearly explains inertia and the difference between centrifugal and centripetal force. You will be able to impress your friends with your new knowledge, and all at a bargain price.

The Author

The author is a retired telecommunications engineering manager, who was employed for almost forty years with a large telecommunications organisation - qualifications consist of a National Certificate in Electrical Engineering and a Full Technological Certificate (Telecommunications) which included a distinction in digital computing at year five; many years ago a member of the Institute of Electrical and Electronic Technician Engineers (IEETE). During his career has worked in and around most power stations in Wales carrying out earth resistivity tests. This included the Nuclear Power Station (now de-commissioned) at Trawsfynydd, and the Ironbridge power stations that occupied a site on the banks of the River Severn at Buildwas in Shropshire, England.

Hobbies include amateur astronomy (member of the Shropshire Astronomical Society), gardening, reading, DIY, travelling, computing, amateur radio (certificate to qualify transmitting and receiving on FM only), a passionate love and appreciation of coastal and country walking.

During his early years as a member of the Youth Hostels Association (YHA) the author completed the YHA (7 peaks) 40 mile marathon walk which entailed climbing the seven highest peaks in South Wales, starting at 0500hrs from the Llanddeusant Youth Hostel in the Black Mountains, Carmarthenshire, crossing and climbing the Brecon Beacons, and completing the marathon the same day at approximately 2200hrs at the George VI Memorial Youth Hostel, Capel-y-Ffin, Black Mountains near Hay-on-Wye.

The author had a motorised boat moored in the Teifi Estuary for a number of years, and when at sea, he and his wife enjoyed the beautiful Welsh coastal scenery including being fascinated by the folding rock strata. When not admiring the views enjoyed mackerel fishing, seal watching and being enchanted by the inquisitive approach and playful 'showing off' by the dolphins in Cardigan Bay.